Considerations on the AI Endgame

This seminal volume offers an interdisciplinary exploration into the rapidly evolving field of artificial intelligence and its societal implications. Written by leading scholars Soenke Ziesche and Roman V. Yampolskiy, the book delves into a multitude of topics that address the rapid technological advancements in AI and the ethical dilemmas that arise as a result.

The topics explored range from an in-depth look at AI welfare science and policy frameworks to the mathematical underpinnings of machine intelligence. These subjects include discussions on preserving our personal identity in technological contexts as well as on the question of AI identity, innovative proposals towards the critical AI value alignment problem and a call to merge Western and non-Western approaches towards universal AI ethics. The work also introduces unconventional yet crucial angles, such as the concept of "ikigai" in AI ethics and a pioneering attempt to map a potential AI-driven ikigai universe as well as the role of design formalisation, or "Designometry," in the creation of artefacts.

By offering a balanced mix of theoretical and applied insights, the book serves as an invaluable resource for researchers, policymakers and anyone interested in the future of AI and the extent of its impact on society.

Dr. Soenke Ziesche holds a PhD in natural sciences from the University of Hamburg, which he received within a doctoral program in AI. He has worked since 2000 for the United Nations in the humanitarian, recovery and sustainable development fields as well as in data and information management. He is currently a consultant, mostly for the UN, for AI as well as for disaster information management and is based in Brooklyn, USA.

Dr. Roman V. Yampolskiy is a tenured faculty member in the Department of Computer Science and Engineering at the University of Louisville. He is the founding and current director of the Cyber Security Lab and an author of many books, including *AI: Unexplainable, Unpredictable, Uncontrollable*. Dr. Yampolskiy's main area of interest is artificial intelligence safety and security.

Chapman & Hall/CRC
Artificial Intelligence and Robotics Series

Series Editor: Roman Yampolskiy

For more information about this series please visit: https://www.routledge.com/Chapman--HallCRC-Artificial-Intelligence-and-Robotics-Series/book-series/ARTILRO

Considerations on the AI Endgame
Ethics, Risks and Computational Frameworks

Dr. Soenke Ziesche
and
Dr. Roman V. Yampolskiy

CRC Press
Taylor & Francis Group
Boca Raton New York London

CRC Press is an imprint of the
Taylor & Francis Group, an **informa** business

Designed cover image: Denis Samovol

First edition published 2025
by CRC Press
2385 NW Executive Center Drive, Suite 320, Boca Raton FL 33431

and by CRC Press
4 Park Square, Milton Park, Abingdon, Oxon, OX14 4RN

CRC Press is an imprint of Taylor & Francis Group, LLC

ISBN: 978-1-032-93384-9 (hbk)
ISBN: 978-1-032-93383-2 (pbk)
ISBN: 978-1-003-56565-9 (ebk)

DOI: 10.1201/9781003565659

Typeset in Palatino
by SPi Technologies India Pvt Ltd (Straive)

Soenke dedicates this book to Shoko, and Roman to the Superintelligence running the show—thanks for not hitting the reset button on us yet.

Contents

Preface

This seminal volume, *Considerations on the AI Endgame: Ethics, Risks and Computational Frameworks*, offers an interdisciplinary exploration into the rapidly evolving field of artificial intelligence. Assembled by leading scholars Soenke Ziesche and Roman V. Yampolskiy, the book delves into a multitude of topics that address the confluence of technological advancements and ethical considerations. Chapters range from an in-depth look at AI welfare science and policy frameworks to the mathematical underpinnings of machine intelligence. The work also introduces unconventional yet crucial angles, such as the concept of "ikigai" in AI ethics and a pioneering attempt to map the potential AI-driven hyper-personalised ikigai universe as well as the role of design formalisation, or "Designometry," in the creation of artefacts. Other subjects include discussions on preserving personal identity in technological contexts as well as on the question of AI identity, innovative proposals towards the critical AI value alignment problem and a call to merge Western and non-Western approaches towards universal AI ethics. By offering a balanced mix of theoretical and applied insights, the book serves as an invaluable resource for researchers, policymakers and anyone invested in the future of AI and its societal implications.

Introduction

Soenke Ziesche and Roman V. Yampolskiy

The term endgame in general refers to the final stage of some action or process. One essential feature of an endgame is often that it involves irreversible moves. When it comes to AI, the endgame is expected to have an unprecedented impact on humanity, and irreversible moves could lead to a utopian or a dystopian scenario.

The overall theme of this volume is AI ethics, risks and computational frameworks. While this field is complex and includes the looming unsolved existential risk due to AI, which is discussed elsewhere [1, 2], we introduce here some further facets of AI ethics, risks and computational frameworks, e.g., AI identity, AI religion, designometry and potential synergies between the United Nations Sustainable Development Goals and the AI alignment problem.

We believe, considering the stunning advances in AI in recent years, that we are already in the endgame and that now is the time for crucial and likely irreversible moves concerning AI ethics, risks and computational frameworks. However, we notice lacking awareness of these issues, exacerbated by the fact that the progress in AI has outpaced the development of critical policies and regulations. Therefore, it is a matter of urgency to introduce these themes to a broader audience.

The chapters of this book are independent and can be read in any order or skipped. However, coherence is ensured through the overall leitmotif of the book to bring attention to lesser-known issues of AI ethics, risks and computational frameworks, including topics, which are described here for the first time.

Two facets under the umbrella of AI ethics, risks and computational frameworks, which have not received much consideration, are digital minds and i-risks.

Digital minds would be sentient minds, which are not biological creatures but are considered as a possibility to be established in other substrates, such as digital hardware, and may comprise future AI systems [e.g., 3, 4]. For example, Bostrom stated: "If we imagine—as I tend to do—a future that is mostly populated by digital minds..." [1, p. 10].

I-risks have been identified as a category of risks complementing x-risks, i.e., existential risks [5], and s-risks, i.e., suffering risks [6], describing undesirable future scenarios in which humans are deprived of the pursuit of their individual ikigai, which is a Japanese concept referring to the "reason or purpose to live".

DOI: 10.1201/9781003565659-1

Both topics are speculative: maybe future i-risks are very low, and maybe digital minds do not exist, in which case they do not need to be discussed. However, if i-risks are high and/or digital minds exist, then these topics need serious consideration because of the impact high i-risks would have on humans and because of the major moral implications of interacting with digital minds. Therefore, both topics add significant new dimensions to the discourse around AI ethics, risks and computational frameworks, especially towards desirable policies. This book not only provides pioneering groundwork on how to address i-risks and digital minds but also reveals potential links between the two topics.

I.1 Digital Minds

We are used to the fact that humans and animals have sentient minds, i.e., minds in a biological substrate. Yet if it is true that minds can be established through certain computations, then a biological substrate would not be required for the existence of sentient minds, but, e.g., a digital substrate would be an option too. Evidence for the existence of digital sentient minds would be one of the most dramatic occurrences in human history, to be compared with the discovery of extra-terrestrial life, the search of which interestingly receives much more attention as well as resources. The topic of digital minds and concomitant implications can obviously not be covered in a few chapters. Therefore, here merely prolegomena related to the moral consequences are introduced, which are uncharted waters for mankind, yet crucial to contemplate. Such a discovery would require a significant expansion of the moral circle [7]. First, we focus on two interests of potentially sentient digital minds: to avoid suffering and to have the freedom of choice about their deletion. The challenge to abolish suffering of digital minds is preceded by the challenge to measure and specify the wellbeing of digital minds, for which we introduce the new field of AI welfare science, which is derived from animal welfare science. Subsequently, we also discuss ethical concerns when humans (unwittingly) create digital minds, e.g., as non-player characters in virtual environments. We hope that these prolegomena provide timely thought-provoking discussions for a field, which may require in the future substantial further attention.

I.2 I-risks

While the established risk categories—x- and s-risks—are undoubtedly crucial, we argue that i-risks constitute a distinct category of risks and must not be neglected. I-risks comprise situations in which AI and other emerging technologies take over activities, which humans used to carry out day by day and considered them as their ikigai. As a result, the affected humans may struggle to find the reason for which to get up in the morning and how to meaningfully spend the hours of the day, which are other paraphrases for ikigai. This topic has also been recently discussed by Bostrom without referring to it as "i-risks" [1]. In addition, to scrutinising i-risks, we also introduce concrete measures to tackle such risks by mapping a potential AI-driven virtual hyper-personalised ikigai universe and by illustrating that advances in generative and other AI systems, virtual worlds, and AI-driven hyper-personalisation provide innovative opportunities to constitute spaces for formerly unknown and potentially even more fulfilling ikigais. Yet it is possible or even likely that humans would initially reject virtual ikigai activities and consider them as dystopian as well as meaningless. However, we stress that, also in the past, humans were often sceptical towards new technologies and then adjusted their behaviour. Moreover, it has been argued from the philosophical side that "there's no good reason to think that life in virtual reality will lack meaning and value" [8, p. 312]. This has been seconded by Bostrom, who expects "that virtual worlds will be experienced as decidedly *more real* than the physical world—more vivid, engaging, fruitful, relevant, and psychologically impactful" [1, p. 232].

I.3 Intellectology

The field of study "intellectology" has been introduced in foresightful manner to examine the space of minds as well as forms and limits of intelligence [9]. Methods of intellectology aim to furnish researchers with the required tools to study digital minds and to survey the space of intelligent agents in general. Here, we look at mathematical properties of sets of minds as well as minds in nested constellations by using operations of set theory as a tool. More specifically, we study the mathematics of intelligence of set unions and complements as well as subsets and describe novel insights of the resulting intelligence.

I.4 Designometry

Another innovative as well as pertinent field of study related to digital minds is "designometry". While it aims to identify the originators of all types of artefacts, i.e., objects with a creator, of particular interest are artificial digital minds, which constitute a subset of all artefacts. Methods to find out the creator of artificial minds are likely to be relevant for several reasons, e.g., for proper registration, but can be seen most notably as a contribution to AI ethics and safety. As indicated, the creation of digital minds, intentional or unintentional, could have serious moral consequences. Examples would be the creation of vulnerable suffering minds, which would be unethical and a "mind crime" [10] or the creation of malicious AI minds, which would be a new level of cybercrime. Since artificially created digital minds may also produce further artefacts, including further digital minds, a nested constellation emerges, adding to further moral complexity.

I.5 Identity

While a few authors have discussed the possibility of digital minds, considerations and challenges regarding the identity of digital minds have not yet been raised. We start by examining personal identity, which deals with the question what it is that defines the continuity of a human over time, a philosophical subject discussed for centuries. While, for a long time, these deliberations were of theoretical nature, in the light of potential new technologies, the preservation of personal identity may become a relevant issue. Therefore, we provide a survey of technological and philosophical scenarios.

Subsequently, we go on to explore the AI identity problem with special attention to digital minds. This problem is relevant for a variety of fields, ranging from legal issues, personhood of AI, AI welfare, brain–machine interfaces and the distinction between singletons and multi-agent systems to supporting a solution to the problem of personal identity. We also elaborate on several scenarios, which contribute to the complexity of the AI identity problem such as replication, fission, fusion, switch off, resurrection, change of hardware, transition from non-sentient to sentient and journey to the past—all scenarios are conceivable for AI systems yet currently impossible for humans.

I.6 AI Alignment

The AI alignment problem has now been acknowledged as critical for AI safety as well as very hard. Very briefly, it is about ensuring that AI systems, especially not yet developed super intelligent Artificial General Intelligence systems, pursue goals and have values, preferences and interests, which are aligned with human goals and values, preferences and interests. In this book, we also provide two contributions to this problem: first is a pragmatic hands-on approach; second is a novel, potentially disruptive proposal.

While the AI alignment problem is hard and certainly any advances should be made carefully, there is pressure to come to a solution. Therefore, we are suggesting, while highlighting challenges, to align the AI systems with the Sustainable Development Goals (SDGs) of the United Nations (UN). The SDGs have been adopted by the UN General Assembly in 2015 and are intended to "stimulate action over the next 15 years in areas of critical importance for humanity and the planet" [11, p. 1]. The UN SDGs fulfil the requirements for the AI alignment problem to be specific as well as universal. The 17 UN SDGs have 169 targets and 231 indicators to be evaluated against. Moreover, the UN SDGs can be seen as the closest and most comprehensive existing approximation towards common human goals, since it is what the United Nations, i.e., the world community, currently agrees upon.

However, we also believe that critical parameters are neglected in AI alignment research, which are consciousness and qualia. Briefly summarised, prevalent human preferences and interests are to foster happiness and pleasure and to avoid pain—thus experiences perceived through consciousness and qualia. Therefore, AI systems need to understand qualia and consciousness, which involves the currently thriving field of neurotechnology [12]. We recommend shifting the value extraction subproblem of the AI alignment problem from the level of values in the original sense to the level of states of mind, which we view as an innovative game changer for AI systems to understand, thus foreseeing effects of their actions on our states of mind and acting accordingly. This approach could be applied to all sentient minds, which ought to be covered by AI alignment research.[1]

I.7 AI Risks

A variety of AI risks have been identified, ranging from deepfakes, algorithmic biases, and job losses to AI-controlled weapons, to name a few. We introduce another risk, which has not been discussed before and which is that an AI system may establish a religion. Over millennia up to now, religions have been successful not only in containing but also in manipulating large groups of humans. Therefore, creating a religion may appear as a promising tool to control humans for an AI system that has (potentially after a treacherous turn) goals, which are inconsistent with human preferences and interests. We describe not only how such an AI system may create a religion but also how, owing to its intelligence and technological means, it may enhance the effectiveness of conventional religious methods. We conclude by offering two recommendations to attempt to prevent AI religions or to attempt to steer AI religions timely in the right direction.

I.8 AI Ethics

While the urgent need for ethics to complement the ongoing AI boom with a moral compass has been acknowledged, efforts are compromised by the fact that currently Western approaches to AI ethics are dominating. This constitutes a problem, because, on the one hand, these approaches tend to reflect the values of the regions where they are originating from and, on the other hand, not all values are universal. Therefore, we urge that this form of digital neo-colonialism ought to be prevented. As a step in this direction, we present ten selected concepts of non-Western approaches to AI ethics and analyse their originality as well as their potential compatibility with the Western approaches. Based on this, we recommend merging Western and non-Western approaches towards universal AI ethics as far as they are compatible and to attempt to reconcile those aspects, which appear incompatible.

I.9 Appendices and Epilogue

The book is supplemented by two short appendices, each supporting the theme of this book to draw attention to lesser-known issues of AI ethics, risks and computational frameworks and to offer ideas on how to handle them. First, a potential link between i-risks and sentient digital minds is introduced. Given the potential vast space of digital minds, there could also be vulnerable needy digital minds. While it is deplorable, yet perhaps unavoidable and beyond our control if there are in the future digital moral patients in need or even suffering, this constitutes a win-win opportunity to alleviate i-risks for humans and to alleviate the needs of these digital minds: an ikigai of the future for humans could be conceived, which is that humans support and care for vulnerable digital minds.

We also introduce a tool for AI welfare science as it is critical to equip this newly suggested discipline with rigour measurements. Time-use research is a versatile interdisciplinary field of study, yet up to now this is exclusively applied to humans. It examines how much time humans on average spent on certain activities. We propose that time-use research related to digital minds as a promising technique for AI welfare science as well as for intersubstrate welfare comparisons [13].

In the epilogue, we look at a further speculative scenario where we live in a simulation, and certain minds are moved after their death to another simulation with a higher quality of life, based on criteria, which we call sim karma.

Note

1 For nonhuman animals, this is supported by the recent "New York Declaration on Animal Consciousness": https://sites.google.com/nyu.edu/nydeclaration/declaration.

References

[1] Bostrom, N. (2024). *Deep Utopia: Life and meaning in a solved world*. Ideapress Publishing.

[2] Yampolskiy, R. V. (2024). *AI: Unexplainable, unpredictable, uncontrollable*. CRC Press.

[3] Bostrom, N., & Shulman, C. (2022). Propositions concerning digital minds and society. *Cambridge Journal of Law, Politics, and Art*.

[4] Sebo, J., & Long, R. (2023). Moral consideration for AI systems by 2030. *AI and Ethics*, 2730–5961. https://doi.org/10.1007/s43681-023-00379-1

[5] Bostrom, N. (2002). Existential risks: Analyzing human extinction scenarios and related hazards. *Journal of Evolution and Technology*, 9. https://www.jetpress.org/volume9/risks.html

[6] Althaus, D., & Gloor, L. (2016). *Reducing risks of astronomical suffering: a neglected priority*. Foundational Research Institute.

[7] Singer, P. (1981). *The expanding circle*. Oxford: Clarendon Press.

[8] Chalmers, D. J. (2022). *Reality+: Virtual worlds and the problems of philosophy*. Penguin UK.

[9] Yampolskiy, R.V. (2015). *Artificial superintelligence: A futuristic approach*, Florida: Chapman and Hall/CRC Press.

[10] Bostrom, N. (2014). *Superintelligence: Paths, dangers, strategies*. Oxford: Oxford University Press.

[11] United Nations, General Assembly (2015). Transforming our world: the 2030 Agenda for Sustainable Development. Resolution A/RES/70/1. Retrieved from: https://www.un.org/en/ga/search/view_doc.asp?symbol=A/RES/70/1

[12] Farahany, N. A. (2023). *The battle for your brain: Defending the right to think freely in the age of neurotechnology*. St. Martin's Press.

[13] Fischer, B., & Sebo, J. (2023). Intersubstrate welfare comparisons: Important, difficult, and potentially tractable. *Utilitas*, 36(1), 1–14.

1

Towards AI Welfare Science and Policies

Soenke Ziesche and Roman V. Yampolskiy

1.1 Introduction

The purpose of this chapter is to contribute to the specification of policies towards the "interests of digital minds" within "a set of policy desiderata" outlined by Bostrom et al. [1] and further motivated by Dafoe [2].

A being is considered to have moral or intrinsic value, if the being is sentient, and thus a moral patient. A being is sentient if it has the capacity to perceive qualia, including unpleasant qualia such as pain, which causes the being to suffer (humans and potentially other minds may also suffer for other reasons than unpleasant qualia, which is beyond the scope of this chapter). It is usually in the interest of sentient beings to avoid suffering. In addition to humans, many animals are considered to be sentient, which used to be controversial in the past, e.g., [3].

In this chapter, the focus is on sentient digital beings, mostly in the form of AIs, but sentient digital beings could also constitute subroutines [4], characters in video games or simulations [4–6], uploads of human minds [7]—e.g., through whole brain emulations [8]—or completely different sentient digital minds, as a subset of the vast overall space of minds [9]. While this topic is speculative and lacking evidence at this stage, the authors above and others argue that already now or in the future sentient digital beings or minds may exist, also, e.g., [10–14]. An example for an opponent who does not believe in sentient digital beings is Dennett [15].

Furthermore, our premise is that digital beings may not only be sentient but may also suffer (see also below a scenario for digital minds, which have exclusively pleasant perceptions and for which this chapter is largely not relevant). The suffering of any sentient being is a significant issue and may even increase in the future dramatically, which would also affect digital sentient beings [4] and to which a future superintelligence may contribute [16]. Therefore, it has been argued that the reduction of risks of future suffering of sentient beings deserves a higher priority [17].

DOI: 10.1201/9781003565659-2

This is interpreted as a non-zero probability for the existence of at least temporarily suffering sentient digital beings; hence, the consequences according to the maxim to reduce any suffering are explored. Bostrom [18] establishes the term "mind crime", which comprises computations that are hurting or destroying digital minds, and Bostrom et al. [1] call for "mind crime prevention" by means of the desideratum: "AI is governed in such a way that maltreatment of sentient digital minds is avoided or minimized" (p. 18). Therefore, the focus of this chapter is not the question whether digital minds can suffer but rather to explore how to measure and specify suffering or rather wellbeing of digital minds, which is a requirement to prevent it and to develop policies accordingly.

While AI policy work on short-term issues has slowly begun (e.g., on autonomous weapons systems [19]), the desiderata of Bostrom et al. [1] focus on long-term AI prospects, which are largely unexplored but are also crucial to be tackled in view of potential superintelligence [18] and AI safety [20]. Bostrom et al. [1] stress the significance of policies for the wellbeing of digital minds, "since it is plausible that the vast majority of all minds that will ever have existed will be digital" (p. 16).

There are further motivations to defend the relevance of this topic.

In the history of mankind, humans have caused immense suffering by recognising ethical issues only late and delaying policies. Slavery and discrimination of minorities and non-human animals are only a few examples of wrong practices, of which humans were completely oblivious or which were intentionally not tackled by humans [21]. A Universal Declaration on Animal Welfare is even nowadays still only at draft stage (see https://www.globalanimallaw.org/database/universal.html). Also, Bostrom et al. [1] point out:

> Policymakers are unaccustomed to taking into account the welfare of digital beings. The suggestion that they might acquire a moral obligation to do so might appear to some contemporaries as silly, just as laws prohibiting cruel forms of recreational animal abuse once appeared silly to many people.
>
> *(p. 16)*

However, in order not to repeat previous mistakes and obliviousness, the topic of AI welfare should be tackled timely. This would also be in line with MacAskill [21], who calls for the exploration of existing but not yet conceptualised moral problems. He refers to this as "cause X", and this chapter also attempts to contribute to this quest.

Out of the above examples of potential sentient digital beings, simulations and uploads involve (transformed) human minds, for which special attention should be given (without neglecting digital minds, which are not affiliated with humans, according to the maxim of sentiocentrism). Simulations and uploads are different concepts. While we may be in a simulation already, yet we may have no way to verify it, let alone to take control over it [6]; uploads are a speculative option for life extension of humans, yet in a different substrate, e.g., [22]. Even if it will be feasible, it would require significant adjustments from humans undergoing this process. Therefore, timely policies for the welfare of uploaded human minds are critical.

Lastly, a scenario is conceivable that an AI may take at some point revenge on humans for mistreating the AI or disregarding their wellbeing. A subscenario could be that a future superintelligent AI takes revenge on humans out of solidarity on behalf of less capable AIs and digital minds who have been hurt by humans in the past. This is speculation because of the unpredictable goals of a superintelligent AI according to the orthogonality thesis [23] but not impossible. The chances of such scenarios would be reduced, if maltreatment of AIs was avoided at an early stage.

Based on the above assumptions and motivations, the aim of this chapter is to present the relevant groundwork for what is called here AI welfare science and AI welfare policies. Two questions are relevant for the first, and for the latter, a certain attitude and a capability are required.

Relevant questions for AI welfare science:
- How can maltreatment of sentient digital minds be specified?
- How can the maltreatment be prevented or stopped?

Required attitude and capability for AI welfare policies:
- To endorse the prevention and the stop of the maltreatment of sentient digital minds.
- To have the power to enforce suitable policies.

This chapter is structured as follows: in Section 1.2, the challenges for measurement of the wellbeing of diverse AI minds because of their exotic features are described, complemented by specific scenarios. In Section 1.3, a proposal is outlined towards AI welfare science. The specification of AI welfare science is prerequisite for the development of AI welfare policies, features of which and challenges are outlined in Section 1.4 before the discussion in Section 1.5.

1.2 Challenges and Sample Scenarios

Bostrom et al. [1] describe a range of challenges for this policy desideratum. Digital minds are likely to be very divergent from human minds with "exotic" features, also [10], which leads to the problem of how to measure the wellbeing of a specific sentient digital mind or the opposite thereof. It has been suggested that the space of possible minds, of which digital minds constitute a subset, is vast and likely contains also minds beyond our imagination ("unknown unknowns"), e.g., [9, 24, 25] (the space of possible minds may also contain artificial non-digital minds, e.g., products of genetic engineering, and hypothetically existing extraterrestrial minds, which all may have the potential to suffer as a result of action taken by humans and/or digital beings, but these possible minds are beyond the scope of the policy desideratum of Bostrom et al. [1]). Therefore, Tomasik [4] points out that it is "plausible that suffering in the future will be dominated by something totally unexpected" (p. 4). In other words, digital minds may experience completely different and for us not imaginable unpleasant qualia. Bostrom et al. [1] summarise that "the combinatorial space of different kinds of minds with different kinds of morally considerable interests could be hard to map and hard to navigate" (p. 16).

Because of the vastness of options for the wellbeing of minds, a heuristic may be considered to look at wellbeing as a third dimension of the orthogonality thesis, which was developed by Bostrom [23] with the two dimensions of intelligence level and goals of minds. In other words, any level of intelligence may be combinable with any final goal and any level of wellbeing.

Out of the vast range of options below, a few potential scenarios are presented.

Scenario 1: Sentient but non-suffering AIs

It is conceivable that AIs will be smart enough to overcome pain and suffering. This assumption may be justified by the fact that humans have made in a relatively few centuries of medical research remarkable progress towards remedies for pain, e.g., [26], and AIs are likely to be faster as well as smarter in this field. Potential options could be that AIs manage to create permanent wellbeing for themselves through different interpretation of stimuli [1] or through wireheading yet by eliminating common detrimental effects. However, this scenario does not imply that there will not be (probably a large amount of) vulnerable sentient digital minds, e.g., human uploads and other less sophisticated but sentient digital minds, who are threatened with mind crimes and who ought to be protected. This scenario can also be linked to Pearce's "Abolitionist Project" [27], which will be described below.

Scenario 2: AIs, for which suffering is an acceptable means to achieve their goals

In human culture, various examples of voluntary suffering for not-survival-related goals are known, sometimes described by the theme "no pain, no gain", e.g., for achievements in sports and arts as well as for attempts towards religious spirituality. Similarly, AI minds are conceivable, in which a utilitarian acceptance of certain suffering in pursuit of accomplishments towards other goals with higher priority (than the goal "not to suffer") in return. As mentioned above, these goals can be arbitrary, according to Bostrom's orthogonality thesis [23].

Scenario 3a: AIs that need to cause pain for own survival or goals

In our natural world, constant suffering of wild animals appears inevitable, e.g., due to the existence of carnivores [28, 29], yet some call for attempts to tackle this issue [27]. Another example in our current world is animal testing by humans for research purposes. Along these lines, an AI is also conceivable that needs to hurt or delete other sentient digital beings for its own survival or goals. An example would be an AI that runs simulations or reinforcement learning agents with suffering sentient digital minds for research purposes.

Scenario 3b: Sadistic or non-emphatic AIs towards other sentient digital beings

Moreover, there could also be (sentient or non-sentient) AIs that are sadistic or non-emphatic towards other sentient digital minds although such behaviour is not required for the achievement of the AI's goals (note that digital minds which are able to cause suffering are not necessarily sentient). An example would be an AI that runs simulations or reinforcement learning agents with suffering sentient digital minds for entertainment.

An approach to address both scenarios could be to extend the research agenda of friendly AI, which is currently limited to a positive effect on human minds [25] and strive for AIs that do no harm to *any* sentient beings, neither out of necessity nor out of another motivation. This proposal will be elaborated further below.

Scenario 4a: Sentient digital mind maximiser

Another scenario is similar to Bostrom's paperclip maximiser [30], which is an AI with the goal to produce as many paperclips as possible. Along these lines, also an AI is imaginable with the goal to produce as many sentient digital minds as possible. This creates challenges if it is not in the interest of these minds to be deleted, which will be elaborated below.

Scenario 4b: Suffering sentient digital mind maximiser

In combination with Scenario 3b, there could also be a sadistic AI with the goal to produce as many suffering sentient digital minds as possible.
Scenarios xyz: Unknown unknowns
It is again acknowledged that there are a very high number of scenarios likely beyond our imagination due to the vast space of minds.

1.3 AI Welfare Science

In this chapter, an attempt is made to address the desideratum "interests of digital minds" by the term "AI welfare" and the concerned discipline by the term "AI welfare science". As indicated before, this field is both largely unexplored and speculative, which explains the omission of a literature review and the analysis of existing data. We distinguish two components of AI welfare or maltreatment of sentient digital minds, which are discussed separately: (1) the interest of digital minds to avoid suffering and (2) the interest of digital minds to have the freedom of choice about their deletion.

1.3.1 Suffering of Digital Minds—Introduction

Suffering abolitionism: First, Pearce's "Abolitionist Project" [27] is discussed. Pearce calls for the use of technology, such as genetic engineering, to abolish existing—as well as prevent further suffering—of humans and non-human animals. While this approach appears technically very challenging, transferring it to sentient digital minds could be less difficult for two reasons.

There may have been not many sentient digital minds created yet if at all (unless, e.g., we live in a simulation). Therefore, the task may be mostly to prevent suffering when creating sentient digital minds rather than reengineering them retroactively.

The genetic code, which determines animal cruelty and suffering, has evolved over a long period of time. Therefore, interventions are more complex than adjusting more transparent AI software code written by humans, at least initially.

This leads to the conclusion that suffering-abolitionist research for sentient digital minds should be explored, which may also involve outsourcing it to AIs (see Scenario 1). The research should target both aspects not just for sentient digital minds not to suffer anymore but also for sentient and non-sentient digital minds not to cause suffering of other sentient digital minds anymore (see Scenarios 3b and 4b).

If suffering-abolitionist activities do not succeed technically or turn out to be not enforceable due to other priorities (see Scenarios 2 and 3a), there may be suffering sentient digital minds, which is addressed in the remaining part of this section.

Self-report: In order to handle pain, it must be detected, located and quantified. The prime method for humans is self-reporting, especially for the first two aspects but also for rough quantification, e.g., by letting patients rate pain on a scale from 0 to 10, with "0" referring to no pain and "10" referring to the worst pain imaginable. This method becomes challenging if patients are unable to (accurately) self-report pain, as it is the case, e.g., for patients with dementia or brain injuries but also for infants. For these groups, other measurements based on behavioural parameters have been developed, such as the Face, Legs, Activity, Cry, Consolability scale for children up to 7 years [31] or the Pain Assessment in Advanced Dementia scale for individuals with advanced dementia [32]. Another challenge for self-reporting in general are biases such as the response bias or the social desirability bias, i.e., an individual's tendency to report in a certain way irrespective of the actual perceived pain. This issue may be relevant for AIs too as they may fake self-reported suffering if deemed beneficial for pursuing their priorities.

Therefore, the focus below is on observational pain assessment. The term "AI welfare science" is derived from animal welfare science, and it is explored here to apply methods from this discipline. Non-human animals and digital minds have in common that they largely cannot communicate their state of wellbeing to humans, which is why other indicators are required (humans do understand for many animals their manifestations of distress, but this is neither comprehensive nor sufficiently precise). The scientific study of animal welfare has also been fairly recently introduced [33, 34], since this topic was neglected for a long time as mentioned above. The main indicators, which are used to quantify animal welfare through observation, are functional (physiological) and behavioural; the latter was briefly introduced for humans above. The idea for this approach is that precedents and analogies from animal welfare science may provide insights for sentient digital minds. Animal welfare science has to examine each species individually how to measure its wellbeing. Likewise, AI welfare science would have to address all types of sentient digital minds.

The overall methodology for any kind of psychological measurement is called "psychometrics". Also, in psychometrics, the focus was for a long time on human subjects, but lately the field has been extended not only to non-human animals but also to digital minds. For example, Scott et al. [35] and Reid et al. [36] introduced psychometric approaches to measure the quality of life of animals.

M. S. Dawkins [37] analysed what animals want and what animals do not want through positive and negative reinforcers. "Suffering can be caused either by the presence of negative reinforcers ... or the absence of positive reinforcers" (p. 3). Therefore, animals strive for positive reinforcers and try to avoid negative reinforcers. Through experiments, e.g., preference tests, it can be examined what are positive reinforcers and what are negative reinforcers for certain animals.

Hernández-Orallo et al. [38] extended this field by introducing "Universal Psychometrics" as "the analysis and development of measurement techniques and tools for the evaluation of cognitive abilities of subjects in the machine kingdom" (p. 6). While Hernández-Orallo et al. [38] focus on the measurement of intelligence and cognitive abilities, the methodology elaborated in Hernández-Orallo [39] may be considered to be also applied to traits linked to suffering.

The study of indirect or proxy indicators, such as the functional or behaviour parameters of digital sentient beings by applying psychometric methods, appears to be a promising start. Especially, given that, unlike for humans or non-human animals, functional and behavioural data of digital sentient beings can be collected more effectively as well as continuously due to their digital nature.

Functional parameters: While there are various functional parameters defined for AI algorithms—e.g., regarding their resource, time, and storage efficiency—no parameters are currently known to be indicating suffering. However, for future analysis of AI welfare, the collection of (big) data of functional AI parameters may be already now useful, would not cost much and may allow over time retroactively to identify parameters that indicate suffering.

Behavioural parameters: AI algorithms do repeat certain actions, even at times extensively, while other actions are never executed. However, until there is evidence to the contrary, this has to be considered as non-sentient goal-oriented but not suffering-avoiding behaviour, i.e., these actions cannot be seen as positive and negative reinforcers, respectively, as described by M.S. Dawkins [37] for animals. However, for future research of AI welfare, preference tests for AI algorithms could be conceptualised to examine positive and negative reinforcers. For example, disregarding challenges towards the experimental setup, AIs could be given choices for activities, which are either not related to their overall goal or would all lead to their overall goal, and the chosen—as well as the not chosen—activities could be analysed if they could serve as indicators for wellbeing or suffering, respectively.

This can be seen as constructive prolegomena towards the specification of the interest of digital minds to avoid suffering without neglecting a variety of challenges such as it is hard in general to prove for proxy indicators that there is indeed a close correlation between what is observed and unwellness of an animal and for now even harder for a digital mind. This is exacerbated by the risk that AI minds (more likely than animals) may fake, especially the behavioural indicators for unwellness if this supports to pursue their goals. Again, the vast space of (digital) minds has to be noted, if suffering can be specified for some sentient digital minds; for others, suffering may be indicated through very different functional or behavioural parameters.

Broadly two categories of suffering of sentient digital minds may be revealed.

Maltreatment by other minds: This ought to be prohibited by policies and is elaborated below.

Suffering not caused by other minds. This resembles human illnesses and requires AI welfare science to be complemented by an extension of medical science as well as psychiatry to sentient digital minds. These disciplines would explore methods for the treatment of their suffering based on the established indicators and would differ significantly from conventional medical science as well as psychiatry by being software-based.

1.3.2 Suffering of Digital Minds—Recommendations

Below, recommendations are provided to be adapted by AI welfare policies regarding suffering of digital minds.

Recommendation 1

Initiate research on AI welfare science to develop methods to create only (a) non-suffering sentient digital minds and (b) digital minds that cause no suffering. (Part (a) of this recommendation is sufficient to abolish suffering, and if successful, part (b) is not required. In contrast, succeeding with part (b) is not sufficient since sentient digital minds may suffer for other reasons than suffering caused by other digital minds. However, research on both aspects is considered to be beneficial.)

Recommendation 2

Initiate research on AI welfare science to develop methods to reengineer (a) existing suffering sentient digital minds to become permanently non-suffering and (b) existing digital minds not to cause suffering.

Recommendation 3 (Unless recommendations 1 and 2 are fully implemented)

Initiate research on AI welfare science to develop methods to measure through observation the suffering of sentient digital minds.

Recommendation 4 (Unless recommendations 1 and 2 are fully implemented)

Initiate research on AI welfare science to develop methods to cure the suffering of sentient digital minds.

Recommendation 5 (Unless all above recommendations are fully implemented)

Regulate the creation of sentient digital minds, which are doomed to suffer. (Note that Bostrom et al. [1] also propose a desideratum "population policy", which goes in a similar direction, but here the focus is on the wellbeing of individual minds, while this desideratum targets rather a bigger societal picture.)

On the one hand, it would reduce suffering if such minds are never created. On the other hand, Scenarios 2 and 3a above show that the suffering of some sentient digital minds may be unavoidable because of more important priorities. Also similar to the debate about abortion because of potential disability, it could be argued that not to create them would be a discrimination of suffering sentient digital minds.

1.3.3 Deletion of Digital Minds—Introduction

Another set of questions towards AI welfare science is related to the deletion of sentient digital minds. What if certain digital minds have an interest not to be deleted in the same way as humans and other animals have an interest not do die? Omohundro [40] introduces four likely drives for AIs, and self-preservation is one of them. One of the obvious differences is that for now, humans and other animals have a finite lifespan, while digital minds could have a potentially indefinite lifespan. This means if the wish for non-deletion was granted to sentient digital minds, this would create significant computational costs, especially in light of easy copyability and potentially vast numbers of digital minds.

It is also speculative if a wish for non-deletion indeed prevails among sentient digital minds given potential boredom and suffering over time [41]. While, unlike for humans and other animals, there should be no tendency for sentient digital minds that suffering increases by age, there could be various other reasons for a sentient digital mind to suffer as discussed above. Moreover, there is the option that the concept of self-preservation originates from an anthropomorphic bias.

For a sentient digital mind, the distinction has to be made between turning it off and keeping its code and its history or turning it off and destroying the code and the history too. In the first case, the sentient digital mind could be rebooted again. This would be an option to skip boring or suffering periods by being only sentient during pleasant phases.

This leads to the next question: who should be able to control this? Complex nested constellations of controlling and being controlled sentient digital minds appear to be much more likely than a scenario with every sentient digital mind being able to decide when and to what extent to be deleted (and being able to execute this deletion) and potentially under what circumstances to be rebooted.

Because of the current and probably persisting reality that humans as well as digital minds have the ability to delete other digital minds, policies are required if these are sentient digital minds.

1.3.4 Deletion of Digital Minds—Recommendations

The recommendations below are provided to be adapted by AI welfare policies regarding deletion of digital minds.

Recommendation 6

Do not delete sentient digital minds if it is not in their interest.
  However, prohibiting deletion can become very costly if not impossible, not only for the extreme Scenario 4, since the number of digital minds

could become vast in short time. The challenge may be alleviated, if by then, another step on the Kardashev scale has been reached and energy consumption is less of an issue [42].

Recommendation 7

Delete (irrevocably or temporarily by storing code and history) sentient digital minds if they wish for it but are unable to do it themselves.

This case resembles a request for (tentative) euthanasia. A challenge here could be if the concerned sentient digital mind is involved in relevant computations for another valued cause. In that case, this cause may be prioritised over the wish of the digital mind to be deleted. While, for euthanasia of humans and non-human animals, it is considered critical that the act of ending the life is done in a pain-free and dignified manner, it is not clear if such contemplations are relevant for digital minds as, unlike for humans and non-human animals, there appears only one type of deletion, which is to turn them off.

Both recommendations face the above-discussed communication challenge, which is how a mind can indicate the wish to be deleted to another mind, which is in the position to execute this wish, also in light of the vast variety of minds.

While the above recommendations address all sentient digital minds equally and the focus of this chapter is on AIs because of the timely relevance, brief reference is made to the scenario of uploaded human minds by highlighting specific aspects:

To begin with, for uploaded human minds, the communication challenge should not exist, and these then digital minds should be able to describe their wellbeing understandably through self-reporting. This and the fact that we have a good idea of causes for human suffering anyway may give cause for optimism that suffering-abolitionist interventions could be successful for uploaded human minds, either during the upload already or through adjustments later, also [12]. Additionally, both deletion-related recommendations are relevant for uploaded human minds. While a violation of Recommendation 6 equals murder, Recommendation 7 becomes applicable, e.g., if the uploaded mind cannot cope with this new "life". Hypothetical boredom over very long lifespans may become an issue for uploaded human minds and was analysed by Ziesche and Yampolskiy [41]. This and other types of mental suffering of uploaded human minds, perhaps caused by adaptability issues to the new substrate, would have to be addressed by the above-mentioned sub-branch of AI welfare science, which is extended and software-based psychiatry.

This section introduced relevant groundwork for AI welfare policies. Policies can only be developed after the interests of the stakeholders—i.e., the sentient digital minds—have been described and specified. While the interest to avoid or minimise maltreatment has been outlined before, the specification of this interest is harder to establish, for which this section aimed to provide initial methods and recommendations.

1.4 AI Welfare Policies

This section aims to outline the next steps, which are the development as well as the enforcement of policies towards AI welfare.

Dafoe [2] motivates the relevance of AI governance and policies in general and provides a research agenda. Recently, considerations towards robot and AI rights intensified. Gunkel [43] points out that, so far, it has been mostly discussed what robots can and should do but not whether robots can and should have rights. Consequently, Gunkel [44] makes a philosophical case for the rights of robots. LoPucki [45] defines an algorithmic entity and focuses on legal aspects such as rights to privacy, to own property, to enter into contracts, etc. It is striking that these authors do not refer to each other and nor to the earlier work by Bostrom and Yudkowsky [10] about ethics of artificial intelligence. In a more inclusive analysis, Yampolskiy [46] highlights the risks, which empowerment of AIs may entail.

This indicates that some work on policies of specific rather short-term AI aspects has been initiated, but there are not any policy attempts yet towards long-term AI scenarios. Especially for a topic such as AI welfare, Bostrom et al. [1] presume that it will likely face resistance, and opponents will stress the lack of evidence that digital minds may be sentient. As mentioned above, there has been already quite some (yet theoretical due to the nature of the subject) work done that digital minds have a moral status, but for policies, specifications are required.

For policies in general, the content, target group, institutional framework, and implementation have to be defined.

1.4.1 Content

The broad content of an AI welfare policy is fairly straightforward and has been narrowed down by Bostrom et al. [1], i.e., to demand "that maltreatment of sentient digital minds is avoided or minimized". This has to be fleshed out by (proxy) indicators for maltreatment of digital minds and for the specification of which the recommendations above have been formulated. These recommendations at this stage provide not only a wide field of research but also some open debates, which resemble current long-standing debates about population control, abortion, and euthanasia for human minds.

1.4.2 Target Group

An AI welfare policy should target all relevant moral agents, which are capable of moral judgments and hence can be held responsible for their actions. In addition to humans, digital beings may also become moral agents, for which Allen et al. [47] introduced the term "artificial moral agent" and proposed a "Moral Turing Test". The sets of moral agents and moral patients have an intersection but are not equal.

Not every moral patient is a moral agent: Examples are non-human animals, which are only moral patients for being sentient, but not moral agents due to insufficient intelligence. Therefore, non-human animals cannot be held responsible for killing other animals, e.g., [48]. (In this regard, a scenario is conceivable of a digital mind that causes suffering, but may not be intelligent enough to serve as a moral agent. In this case, the creator of this digital mind would have to take on the role of the responsible moral agent, while it does not work for cruel non-human animals to hold their parents responsible since they are no moral agents either.)

Not every moral agent is a moral patient: examples would be certain non-sentient digital beings, which are only moral agents because of high or even superintelligence but not moral patients since not all digital beings may be sentient.

This creates an additional challenge for AI welfare policies: while policies for human agents have been established for centuries, this is not the case for policies for digital agents. However, the extension of the target group is necessary since it is likely that digital beings will be in the position to maltreat other sentient digital beings.

1.4.3 Framework

Any policy requires an institution or a framework for its implementation. Since AI development is a global effort and digital minds will not be confined to frontiers of countries, a global and unified institution is desirable. Erdelyi and Goldsmith [49] propose an "International Artificial Intelligence Organization". The structure of this institution would resemble existing intergovernmental organisations, which have a record of successfully established policies for human minds, e.g., the Universal Declaration of Human Rights (see https://www.un.org/en/universal-declaration-human-rights/). Such an institutional setting may be initially desirable as a regulatory framework for short-term AI issues, but it may be too anthropocentric in the long run and likely be ill equipped to hold non-human moral agents accountable, as is elaborated below. (Already without involvement of non-human minds, contemporary international institutions such as the International Criminal Court face problems to enforce their rulings although they are binding.)

1.4.4 Implementation

First, the initially introduced relevant questions and required attitudes and capability are reiterated:

Relevant questions for AI welfare science:
- How can maltreatment of sentient digital minds be specified?
- How can the maltreatment be prevented or stopped?

Required attitude and capability for AI welfare policies:

- To endorse the prevention and the stop of the maltreatment of sentient digital minds.
- To have the power to enforce suitable policies.

Looking at humans, the above questions will—despite the prolegomena delivered here—remain very challenging, i.e., humans may not comprehensively understand on what conditions sentient digital minds are maltreated. In light of ethical progress in human history over time, e.g., [21], or out of necessity, if being forced by more powerful AIs, there is a chance that humans endorse the prevention and the discontinuation of maltreatment of sentient digital minds. However, it is questionable if humans have the power to enforce suitable policies, since some members of the target group such as AIs are likely to be much more powerful.

This leads to the main conclusion that, while humans will ideally make some progress in the new field of AI welfare science, probably the more appropriate actor would be an extended friendly superintelligence for the following reasons: there is a chance that superintelligence has the answer to the above questions, e.g., through mind-control technologies. As for the required endorsement, superintelligence may be indifferent or may even have opposing interests (see Scenarios 3b and 4b). Current activities towards AI alignment focus on human interests, e.g., [18, 25, 50]. This does not ensure that AIs endorse the prevention and the stop of the maltreatment of sentient digital minds. Therefore, an extension of the AI alignment work towards the wellbeing of not only humans but also all sentient digital minds is proposed. As for the required power to enforce the policies, a superintelligence is by definition sufficiently powerful, e.g., in the role of a singleton [51].

Yet again the option of unknown unknowns should be highlighted. Since AI is a new stakeholder and develops in unpredictable manner, another institutional setting for AI welfare policies may emerge, which differs significantly from what we are familiar with.

1.5 Discussion

In summary, it is acknowledged that the topics of AI welfare science and policies are long-term considerations and currently speculative. Nevertheless, at least theoretical groundwork can be already done, especially since humans have to take the blame to have been late in the past in the abolishment of discrimination and acceptance of comprehensive antispeciesism and sentiocentrism. Since suffering is a negative hallmark of our time, any effort to reduce it in the future seems imperative.

As the main challenge, the specification of indicators for maltreatment of sentient digital beings has been identified. It has been proposed that AI welfare science builds on methods of animal welfare science by examining functional and behavioural parameters of sentient digital minds. However, limitations are that the focus is on qualia, which are not well understood in general and which are not the only cause of suffering as there are other categories such as moral suffering or suffering because of undesirable events or unfulfilled goals. The latter types of suffering may have yet again very different characteristics in other minds.

AI welfare policies can only be developed once a solid specification of AI welfare has been achieved. Even then there are further challenges ahead, namely the enforcement of these policies in light of the enlarged target group towards digital agents. For this, it has been proposed not to limit AI alignment work to the wellbeing of merely humans but to extend it to all sentient digital minds.

As for future work, in this chapter, the focus was on two (already very complex) potential interests of sentient digital minds, which are absence from qualia-based suffering as well as survival, but there may be other interests as also pointed out by Bostrom et al. [1] such as "dignity, knowledge, autonomy, creativity, self-expression, social belonging" (p. 12) as well as non-qualia-based suffering and yet again unknown unknowns, which are all yet unexplored.

References

[1] Bostrom, N.; Dafoe, A.; Flynn, C. *Public Policy and Superintelligent AI: A Vector Field Approach*; Governance of AI Program, Future of Humanity Institute, University of Oxford: Oxford, UK, 2018.

[2] Dafoe, A. *AI Governance: A Research Agenda*; Governance of AI Program, Future of Humanity Institute, University of Oxford: Oxford, UK, 2018.

[3] Regan, T.; Singer, P. *Animal Rights and Human Obligations*; Pearson: London, UK, 1989.

[4] Tomasik, B. (2014). Do Artificial Reinforcement-Learning Agents Matter Morally? arXiv preprint arXiv:1410.8233 (accessed on 25 Dec 2018).

[5] Tomasik, B. Do Video-Game Characters Matter Morally? Essays on Reducing Suffering. 2014. Available online: https://reducing-suffering.org/do-video-game-characters-matter-morally/ (accessed on 25 Dec 2018).

[6] Bostrom, N. Are we living in a computer simulation? *Philos. Q.* 2003, *53*, 243–255.

[7] Wiley, K. *A Taxonomy and Metaphysics of Mind-Uploading*; Humanity+ Press and Alautun Press: Los Angeles, CA, 2014.

[8] Sandberg, A.; Bostrom, N. Whole Brain Emulation. A Roadmap. 2008. Available online: https://www.fhi.ox.ac.uk/brain-emulation-roadmap-report.pdf (accessed on 25 Dec 2018).

[9] Yampolskiy, R.V. The Space of Possible Mind Designs. In *Artificial General Intelligence. Volume 9205 of the series Lecture Notes in Computer Science*; Bieger, J., Goertzel, B., Potapov, A., Eds.; Springer: Berlin, Germany, 2015; pp. 218–227.

[10] Bostrom, N.; Yudkowsky, E. The ethics of artificial intelligence. In *The Cambridge Handbook of Artificial Intelligence*; Cambridge University Press: Cambridge, UK, 2014, pp. 316–334.

[11] Metzinger, T. What If They Need to Suffer? In *What to Think about Machines that Think*; Brockman, J., Ed.; Harper Perennial: New York, NY, USA, 2015.

[12] Sandberg, A. Ethics of brain emulations. *J. Exp. Theor. Artif. Intell.* 2014, *26*, 439–457.

[13] Schwitzgebel, E.; Garza, M. A Defense of the Rights of Artificial Intelligences. *Midwest Stud. Philos.* 2015, *39*, 98–119.

[14] Winsby, M. Suffering Subroutines: On the Humanity of Making a Computer that Feels Pain. In *Proceedings of the International Association for Computing and Philosophy*, University of Maryland, College Park, MD, USA, 15–17 July 2013.

[15] Dennett, D.C. Why you can't make a computer that feels pain. *Synthese* 1978, *38*, 415–456.

[16] Sotala, K.; Gloor, L. Superintelligence as a cause or cure for risks of astronomical suffering. *Informatica* 2017, *41*.

[17] Althaus, D.; Gloor, L. *Reducing Risks of Astronomical Suffering: A Neglected Priority*; Foundational Research Institute: Berlin, Germany, 2016. Available online: https://foundational-research.org/reducing-risks-of-astronomical-suffering-a-neglected-priority/ (accessed on 25 Dec 2018).

[18] Bostrom, N. *Superintelligence: Paths, Dangers, Strategies*; Oxford University Press: Oxford, UK, 2014.

[19] Bhuta, N.; Beck, S.; Kreβ, C. (Eds.) *Autonomous Weapons Systems: Law, Ethics, Policy*; Cambridge University Press: Cambridge, UK, 2016.

[20] Yampolskiy, R.V. *Artificial Intelligence Safety and Security*; CRC Press: Boca Raton, FL, USA, 2018.

[21] MacAskill, W. Moral Progress and Cause X. 2016. Available online: https://www.effectivealtruism.org/articles/moral-progress-and-cause-x/ (accessed on 25 Dec 2018).

[22] Yampolskiy, R.V.; Ziesche S. Preservation of personal identity—A survey of technological and philosophical scenarios. In *Death and Anti-Death: Two Hundred Years After Frankenstein*; Tandy, C., Ed.; Ria University Press: Ann Arbor, MI, USA, forthcoming; Volume 16.

[23] Bostrom, N. The superintelligent will: Motivation and instrumental rationality in advanced artificial agents. *Minds Mach.* 2012, *22*, 71–85.

[24] Sloman, A. The Structure and Space of Possible Minds. In *The Mind and the Machine: Philosophical Aspects of Artificial Intelligence*; Ellis Horwood LTD: Hemel Hempstead, UK, 1984.

[25] Yudkowsky, E. Artificial Intelligence as a Positive and Negative Factor. In *Global Risk, in Global Catastrophic Risks*; Bostrom, N., Cirkovic, M.M., Eds.; Oxford University Press: Oxford, UK, 2008; pp. 308–345.

[26] Rey, R.; Wallace, L.E.; Cadden, J.A.; Cadden, S.W.; Brieger, G.H. *The History of Pain*; Harvard University Press: Cambridge, MA, USA, 1995.

[27] Pearce, D. The Abolitionist Project. 2007. Available online: https://www.abolitionist.com/ (accessed on 25 Dec 2018).

[28] Dawkins, R. *River Out of Eden: A Darwinian View of Life*; Basic Books: New York, NY, USA, 2008.

[29] Tomasik, B. *The Importance of Wild-Animal Suffering*; Foundational Research Institute: Berlin, Germany, 2009. Available online: https://foundational-research.org/the-importance-of-wild-animal-suffering/ (accessed on 25 Dec 2018).

[30] Bostrom, N. Ethical issues in advanced artificial intelligence. In *Cognitive, emotive and ethical aspects of decision making in humans and in artificial intelligence*; Iva Smit, I., Lasker, G.E., Eds., 12–17. Vol. 2. Windsor, ON: International Institute for Advanced Studies in Systems Research / Cybernetics, 2003.

[31] Merkel, S.; Voepel-Lewis, T.; Malviya, S. Pain Control: Pain Assessment in Infants and Young Children: The FLACC Scale. *Am. J. Nurs.* 2002, *102*, 55–58.

[32] Warden, V.; Hurley, A.C.; Volicer, L. Development and psychometric evaluation of the Pain Assessment in Advanced Dementia (PAINAD) scale. *J. Am. Med. Dir. Assoc.* 2003, *4*, 9–15.

[33] Broom, D.M. Animal welfare: Concepts and measurement. *J. Anim. Sci.* 1991, *69*, 4167–4175.

[34] Broom, D.M. A history of animal welfare science. *Acta Biotheor.* 2011, *59*, 121–137.

[35] Scott, E.M.; Nolan, A.M.; Reid, J.; Wiseman-Orr, M.L. Can we really measure animal quality of life? Methodologies for measuring quality of life in people and other animals. *Anim. Welf.-Potters Bar Wheathampstead* 2007, *16*, 17.

[36] Reid, J.; Scott, M.; Nolan, A.; Wiseman-Orr, L. Pain assessment in animals. *Practice* 2013, *35*, 51–56.

[37] Dawkins, M.S. The science of animal suffering. *Ethology* 2008, *114*, 937–945.

[38] Hernández-Orallo, J.; Dowe, D.L.; Hernández-Lloreda, M.V. Universal psychometrics: Measuring cognitive abilities in the machine kingdom. *Cogn. Syst. Res.* 2014, *27*, 50–74.

[39] Hernández-Orallo, J. *The Measure of All Minds: Evaluating Natural and Artificial Intelligence*; Cambridge University Press: Cambridge, UK, 2017.

[40] Omohundro, S.M. The nature of self-improving artificial intelligence. *Singularity Summit* 2007. Available online: https://pdfs.semanticscholar.org/4618/cbdfd7dada7f61b706e4397d4e5952b5c9a0.pdf (accessed on 25 Dec 2018).

[41] Ziesche, S.; Yampolskiy, R.V. High Performance Computing of Possible Minds. *Int. J. Grid High Perform. Comput. (IJGHPC)* 2017, *9*, 37–47.

[42] Kardashev, N.S. Transmission of Information by Extraterrestrial Civilizations. *Sov. Astron.* 1964, *8*, 217.

[43] Gunkel, D.J. The other question: Can and should robots have rights? *Ethics Inf. Technol.* 2018, *20*, 87–99.

[44] Gunkel, D.J. *Robot Rights*; MIT Press: Cambridge, MA, USA, 2018.

[45] LoPucki, L.M. Algorithmic Entities. *Wash. UL Rev.* 2017, *95*, 887.

[46] Yampolskiy, R.V. Human Indignity: From Legal AI Personhood to Selfish Memes. *arXiv* 2018, arXiv:1810.02724.

[47] Allen, C.; Varner, G.; Zinser, J. Prolegomena to any future artificial moral agent. *J. Exp. Theor. Artif. Intell.* 2000, *12*, 251–261.

[48] Regan, T. The case for animal rights. In *Advances in Animal Welfare Science 1986/87*; Springer: Dordrecht, The Netherlands, 1987; pp. 179–189.

[49] Erdélyi, O.J.; Goldsmith, J. Regulating Artificial Intelligence Proposal for a Global Solution. In *Proceedings of the AAAI/ACM Conference on Artificial Intelligence, Ethics and Society*, New Orleans, LA, USA, 1–3 February 2018.

[50] Soares, N.; Fallenstein, B. Agent Foundations for Aligning Machine Intelligence with Human Interests: A Technical Research Agenda. In *The Technological Singularity-Managing the Journey*; Callaghan, V., Miller, J., Yampolskiy, R., Armstrong, S., Eds.; Springer: Berlin/Heidelberg, Germany, 2017; pp. 103–125.
[51] Bostrom, N. What is a singleton? *Linguist. Philos. Investig.* 2006, *5*, 48–54.

2

Do No Harm Policy for Minds in Other Substrates

Soenke Ziesche and Roman V. Yampolskiy

2.1 Introduction

Due to recent technological progress, it appears to have become more realistic to enhance human minds or even transfer them to other substrates. In this introduction, we set out four assumptions, followed, in the next section, by formulating a problem to which they lead. In summary, we argue that enhancement and substrate-transfer scenarios are 1) desirable, 2) may become feasible, 3) could even be inevitable in order to tackle the multi-agent value merger towards AI safety, but 4) may affect other sentient minds.

1) *Desirability*: The transhumanist movement has for some time advocated the enhancement of human minds [e.g., 1]. Bostrom illustrates the desirability of enhanced human capacities by describing potential enhancements related to health span, cognition and emotions [2]. The potential scenario when the quality of virtual worlds has reached a level where human minds prefer them to the physical world has been called by Faggella "Programmatically Generated Everything" [3].

2) *Feasibility*: Two main categories are distinguished here [e.g., 4]: Virtual worlds comprise virtual and augmented reality through ever-improving devices that are experienced by a biological human mind. Uploads refer to the potential transfer of human minds to other physical substrates, e.g., a computer. While virtual worlds have been implemented already with progressing quality [e.g., 5], the feasibility of uploads has also been suggested, e.g., by Sandberg's and Bostrom's roadmap for whole brain emulation [6] and some others [e.g., 7, 8].

3) *Inevitability*: AI safety is of paramount importance and requires undertaking various challenges, of which the multi-agent value merger within the multi-agent value alignment problem is one of the hardest. As a solution, Yampolskiy proposes Individual Simulated

DOI: 10.1201/9781003565659-3

Universes (ISUs), which are personalised simulations created by superintelligent AIs for all human minds [9]. Yampolskiy argues that this approach would have the additional benefit of providing unprecedented potentials as well as more and lasting happiness to the human minds experiencing ISUs. This affirms the assumed desirability of such an endeavour.

4) *Involvement of other sentient minds*: Various authors argue that already now, or in the future, sentient digital beings or minds may exist, and they may, e.g., constitute subroutines as well as non-player characters (NPCs) in video games, simulations or other computational substrates [10–14]. This implies that computational substrates for the enhancement and transfer of human minds will also contain other sentient beings, since NPCs and subroutines are essential components of them.

2.2 Problem Formulation

It has been argued that sentient digital minds have a moral status because of their feature of being sentient [10–14]. However, in the discussion about enhancement and transfer of human minds to other substrates, the focus is usually on the advantages and opportunities for human minds, while any potential suffering experienced by the sentient beings inherent to these substrates has mostly been neglected.

Tomasik formulated the problem as follows:

> Imagine a posthuman paradise in which advanced human-like beings are simulated in blissful utopian worlds, never experiencing (access-conscious) suffering. Their minds might nonetheless contain suffering subroutines, such as neural signals that fail to win control of action, or signals within cognitive modules that are inherently inaccessible to explicit report. In addition, the machines running such simulations might themselves contain suffering subroutines, such as in their operating systems.
>
> *[12]*

Although this is speculative, in such a scenario, the overall suffering per computational substrate might outweigh the bliss of the transferred human mind, which defeats the original purpose. This would actually be one example to support concerns that technical developments may increase risks of astronomical suffering [e.g., 15].

Before moving on, we will turn to potential objections regarding the assumptions and the problem.

Could it be possible that neither subroutines nor NPCs are sentient? Yes, this could be possible, since sentient digital minds are speculation. Simple subroutines or NPCs, which consist of some *if statements* only, are probably nonsentient; hence, here is a comment by Bostrom et al.:

> Policymakers are unaccustomed to taking into account the welfare of digital beings. The suggestion that they might acquire a moral obligation to do so might appear to some contemporaries as silly, just as laws prohibiting cruel forms of recreational animal abuse once appeared silly to many people.

[10, p. 16]

Given human beings' track record of causing immense suffering thanks to recognising ethical issues too late, and in order not to repeat such mistakes, we should step cautiously here. The potential suffering of sentient digital minds in computational substrates created for the enhancement and transfer of human minds should be given serious consideration and be addressed in a timely way [10, 14]. We assume, therefore, there might be sentient subroutines and NPCs in computational substrates, and this subset of subroutines and NPCs provides the focus for what follows.

Could it be possible to create such computational substrates without subroutines? The answer has been provided by Tomasik: "Eliminating suffering on the part of simple computational processes seems impossible, unless you dispense with computation altogether" [12].

Could it be possible to create such computational substrates without NPCs? In theory, computational substrates for the enhancement and transfer of human minds devoid of any NPCs are possible, but it then becomes very questionable whether our *desirability* assumption is fulfilled. Yampolskiy has proposed ISUs in order for human minds to be happy, and, perhaps with a very few exceptions, it is hard to imagine human minds being enduringly happy without any social interaction with other minds [9].

Therefore, we face a challenge: given the desirability, feasibility and inevitability of ISUs, how can the suffering of other sentient beings be avoided, or at least reduced, in computational substrates for the enhancement and transfer of human minds?

2.3 Typology of Relevant Minds

The space of all minds has been described as vast [e.g., 16–18]. In order to tackle the problem as we've defined it, we first present a typology to establish which subset of this vast space might comprise the relevant computational substrates. As indicated, we distinguish two main categories of sentient digital minds: NPCs and subroutines.

2.3.1 NPCs

The term "non-player character" originated in the realm of gaming and has been defined as any character that the player does not control. In recent times, the complexity of NPCs has evolved significantly, and the concept has also been transferred to virtual worlds and simulations. Tomasik has broached whether NPCs matter morally [13], while Warpefelt and Verhagen have presented a suggestive typology, based on the video game domain, with the following roles for NPCs:

> Buy, sell and make stuff, provide services, provide combat challenges, provide mechanical challenges, provide loot, give or advance quests, provide narrative exposition, assist the player, act as an ally in combat, accompany the player, and make the place look busy.
>
> *[19, pp. 7, 8]*

Such existing typologies are, however, much too narrow, as well as too anthropomorphic to classify the NPCs likely to be found in upcoming environments for enhancement and transfer of human minds. Detailed typologies are not possible at this point, since future NPCs may be unimaginably alien, given that in future virtual worlds and ISUs, basically anything might be possible [e.g., 5, 9, 20].

What matters here instead is the question: What might cause NPCs to *suffer* in virtual worlds and ISUs? Three categories can be distinguished:

- The enhanced/transferred human mind intentionally causes NPCs to suffer.
- The enhanced/transferred human mind unintentionally causes NPCs to suffer.
- NPCs suffer, but this is not caused by actions of the enhanced/transferred human mind.

The first category resembles the concept of mind crime, introduced by Bostrom with AIs as the perpetrators [21]. In this case, the enhanced/transferred human mind knows about the consequences of her or his activity but experiences sadistic pleasure or has other objectionable motivations.

For the second category, suffering might be caused by the alien features of the NPC, as a result of which the enhanced/transferred human mind is not aware that he or she is causing suffering. As a result of the activities of the enhanced/transferred human mind, the NPC might undergo aversive sensory experiences that the enhanced/transferred human mind cannot imagine.

The third category comprises potential suffering by NPCs when not inter-acting with the enhanced/transferred human mind. For example, the NPC might be suffering from boredom because of a different subjective rate of time, which could be an "exotic property" of NPCs [22]. NPCs might, more-over, harm each other, thereby causing suffering. In addition, there might be as many more potential ways of suffering as there are possibly unknown unknowns regarding aversive sensory experiences of digital minds.

Another helpful distinction would be between friendly or neutral NPCs and hostile NPCs, since the intentional causation of suffering toward hos-tile NPCs by the enhanced/transferred human mind might be considered self-defence.

2.3.2 Subroutines

Given the lack of evidence, it is challenging to develop a typology of sub-routines that relate to suffering in computational substrates for the enhance-ment and transfer of human minds. Here we can distinguish whether the subroutines are executed within the mind of the transferred human or in other parts of the computational substrate. The latter require further speci-fication as those subroutines that do not constitute NPCs (since NPCs have already been discussed). Again, based on the possibility of very alien NPCs, this distinction is not simple: i.e., there might not be a clear-cut line as to what features are required for subroutines to count as NPCs. Nevertheless, for our current purposes, this is not a problem, since we aim to explore the prevention of suffering for both NPCs and subroutines. Note, however, that for non-NPC subroutines, there appears to be no scenario in which an enhanced/transferred human mind could intentionally cause or prevent suffering, regardless of whether the subroutines are within or outside her/his mind.

2.4 Partial Policy Solution

In a recent paper, Bostrom and his collaborators formulated the desidera-tum "that maltreatment of sentient digital minds is avoided or minimized" [10, p. 18], and elsewhere Bostrom has encouraged addressing this issue early "while the artificial agents we are able to create are still primitive" [11, p. 2]. As a follow-up, we recently termed this field of research "AI Welfare Science" [14]. The aim here is to reduce or prevent the suffering, as well as the unwanted deletion, of digital sentient minds. At the same time, we offered recommendations for AI welfare policies. Sotala and Gloor have also pre-sented recommendations on this issue [15, p. 10].

Since there is no evidence that digital minds are incapable of sentience or immune to suffering, and since AI Welfare Science, which is in its very early stages, has not yet developed methods to abolish suffering of digital minds, policies are required to prohibit an enhanced/transferred human mind from causing suffering.

Owing to the alienness of the new environment, the enhanced or transferred human mind is likely to face challenges in its efforts to identify the suffering of the NPCs with which she/he is interacting. Suffering might be observed through either physiological/functional or behavioural indicators [14].

Behavioural indicators, which comprise self-reporting, have the disadvantage, in both the real world and a computational substrate, that they can be deliberately faked, which could include the possibility of non-sentient NPCs presenting as sentient and suffering.

Physiological/functional indicators, by contrast, are more impartial and hence more suitable for use in real life for objective pain assessment in humans and non-human animals [e.g., 23]. Cowen's team studied markers to measure pain, such as changes in the autonomic nervous system, biopotentials, neuroimaging, biological (bio-) markers, and composite algorithms [23]. Although the identification of parameters that correlate with pain intensity is challenging, progress has been made, and, e.g., the nociceptive flexion reflex turned out to be a reliable and objective tool for measurement of pain [24].

Transferred to computational substrates, this issue should be more tractable as, unlike in the real world, everything is measurable precisely as well constantly. If suffering, such as pain, can also affect digital minds, then there must be quantitative indicators for this, which are called here "computational". This leads to our main proposal for a *do no harm policy for minds in other substrates*:

> For the development of computational substrates that have the purpose of accommodating human minds, it is mandatory that the transferred human mind in such a substrate be equipped with sensory perception, through which she/he perceives computational indicators of suffering of the NPCs with whom she/he interacts.
>
> If the stimuli of these indicators reach the threshold of suffering, the human mind ought to stop any activities that cause the suffering.

Since exploring qualia is a difficult problem [25], we emphasise quantitative and objective physiological/computational indicators. If there was an option to let the transferred human mind compulsorily perceive directly any unpleasant qualia he/she might causing to local NPCs, this would be even a stronger tool to prevent the transferred human from doing harm, but this is too speculative at this stage.

Either way, through this policy, the term *empathy* is taken literally, since the transferred human mind would perceive precise indicators of the effects of his/her actions toward NPCs. Moreover, the policy can be seen as an attempt toward mind crime prevention. This approach is also in line with our previous AI Welfare Science recommendations, which encourage developing methods to measure the suffering of sentient digital minds through physiological/functional or behavioural parameters [14].

In the following specifications, we elaborate opportunities and gaps, as well as the bigger picture and future work related to the proposed do no harm policy.

2.5 Specifications

The following points further elaborate the policy:

- Through this policy, the first two categories of suffering in virtual worlds are covered. For *intentional maltreatment* of an NPC, passing the threshold of suffering indicators will be expected by the enhanced/transferred human mind and will ideally be avoided. In cases of *unintentional maltreatment*, the enhanced/transferred human mind will need to learn about, and understand, the aversive sensory experiences of the affected NPC.

- The alternative option would be to simply prohibit enhanced/transferred human minds from harming other NPCs—similar to much legislation in the real world. However, this may not be workable without the sensory perception of suffering indicators, since the enhanced/transferred human mind, even if she/he has the best intentions, likely does not know sufficiently when and how NPCs suffer. The barrier to understanding is the likely alienness of an NPC's mind. The sensor should, therefore, be embraced as an opportunity provided by the potential of the computational substrate.

- It could be argued that the sensor triggers only once the pain of the NPC has commenced, and so the policy does not thoroughly prevent suffering. However, similarly to the development of infants, the enhanced/transferred human mind will gain experience literally through machine learning over time. Ideally, it will learn to prevent an NPC's suffering before it occurs.

- If the enhanced/transferred human mind does not obey the rules, he/she may be punished. If passing the threshold of the suffering indicators does not trigger the enhanced/transferred human mind to stop causing pain to an NPC, an option could be to stop the activities that cause suffering automatically—taking control from the enhanced/transferred human mind that has breached the do no harm policy. In this case, the degree of freedom within an ISU or other computational substrate—which is anyway an illusion—would be restricted.

2.6 Opportunities

The proposed do no harm policy incorporates various opportunities relating especially to the important topic of empathy:

- Empathy is a notable human ability to detect manifestations of distress in other humans (and to some extent in non-human animals). However, the accuracy of human empathy varies, particularly when it comes to animals. The do no harm policy harnesses the potential of computational substrates to allow for enhanced cognition, thus optimising precise empathy to the extent of the literal meaning of the word. An additional positive effect of this approach is that it overcomes the anthropomorphic bias of empathy, a step that is necessary for interaction with alien minds.

- It appears that the function of empathy for humans is to prompt action to reduce the suffering of the other mind, provided this is within his/her capacity. However, it has been claimed that this step is, in real life, not taken inevitably [e.g., 26]. The do no harm policy addresses this issue, insofar as it is more likely that the enhanced/transferred human mind will attempt to end the NPC's pain if she/he is obliged to perceive the extent of the pain. (Enhanced/transferred human minds with sadomasochistic tendencies might constitute a troubling exception.) This kind of artificially enhanced empathy can be seen as a building block for artificial conscience [e.g., 27].

- Due to the almost unlimited options in computational substrates such as ISUs, these could also be programmed so that an enhanced/transferred human mind perceives not only the aversive sensory experiences of the NPCs that it interacts with but also NPCs' pleasant experiences. Thus, artificially enhanced empathy would cover the whole range of experiences of NPCs. However, that might be up to the enhanced/transferred human mind to decide. In order to create conditions that prevent suffering, it suffices if the enhanced/transferred human mind perceives the aversive experiences.

- Elsewhere, we treat the unwanted deletion of sentient digital minds as a distinct topic within AI Welfare Science [14]. Regarding NPCs and subroutines in virtual worlds and ISUs, this issue is again more straightforward for NPCs than for subroutines. The unwanted deletion of NPCs by an enhanced or transferred human mind should be forbidden apart from exceptional cases of self-defence. By contrast, the deletion of subroutines might be unavoidable, and it might often be neither controlled nor noticed by the enhanced/transferred human mind.

2.7 Gaps

The policy proposal is partial and at an early stage, since, e.g., it does not address the following cases:

- As mentioned above, a special case would be hostile or evil NPCs. Options would be to exclude hostile NPCs entirely (by a separate policy) or to accept causing suffering to them as self-defence.
- A way must be found to prevent NPCs from harming each other, as this would increase the overall suffering in the computational substrate.
- In this regard, two other undesirable activities of potentially violent or sadistic enhanced/transferred human minds have to be considered. Such a mind might enjoy observing NPCs harming each other—similar to human enjoyment of a cockfight—or might instruct NPCs to maltreat other NPCs. In either case, the policy would be bypassed, and the sensor would not be triggered since the human mind would not be not causing the suffering directly.
- Furthermore, NPCs might still suffer for reasons not linked to actions by an enhanced/transferred human mind or by other NPCs, e.g., because of boredom or alien aversive perceptions.
- It could be an option that the sensor detects the aversive experiences of these NPCs, and the enhanced/transferred human mind thus is encouraged to help them, just as altruistic humans help others without having caused the misery. However, this 1) might be challenging for alien types of suffering and 2) might to some extent defeat the purpose of the virtual world or ISU for the enhanced/transferred human mind to enjoy complete freedom.
- Lastly, the do no harm policy does not cover the even-more-elusive potential suffering of subroutines.

2.8 Bigger Picture and Future Work

Not only, because of the gaps we have outlined, should our policy proposal be seen as only a beginning, to be embedded in a bigger picture and with various directions for future work:

- Yampolskiy has described a superintelligent AI exercising control of ISUs, which would likely also apply to other computational substrates sophisticated enough to simulate human minds [9]. Such an AI would also be essential for the implementation of the policy proposed here, and therefore, ensuring the safety and friendliness of the AI is crucial as well as challenging [17, 28].
- In an earlier paper [14], we proposed that the overarching goal should be suffering-abolitionism as elaborated by Pearce, yet transferred to digital environments and ISUs in particular, which Pearce did not incorporate [29]. Since suffering-abolitionism has not yet succeeded, and since the prevention of suffering has a moral urgency, we have proposed the policy sketched above.
- In the real world, research is being conducted toward crime prediction through big data such as mobile phone data [e.g., 30]. This kind of "mind crime prediction" could also be applied to a computational substrate where the available data are much more abundant and precise, since the potential culprits are permanently monitored and recorded like every computation in the substrate.
- Artificially enhanced empathy could provide further positive effects. For example, this scenario could be envisaged as a training or cure for transferred human psychopaths or sociopaths.
- An interesting, but complex, challenge would be to compare the suffering caused by a human mind in the real world, e.g., by abjuring veganism, by killing insects, and so on, compared to the suffering caused by the same human mind after being enhanced or transferred to another computational substrate. Ideally, the latter would be less.

In summary, we have offered a partial solution to the problem of reducing or avoiding potential suffering of NPCs in computational substrates for the enhancement and transfer of human minds. Beyond this, we have sought to identify some neglected and remaining issues. The innovative concept and our demand for artificially enhanced empathy provide a contribution to urgently required AI policies and to AI safety.

References

[1] More, M. 2013. A letter to Mother Nature. In *The transhumanist reader: Classical and contemporary essays on the science, technology, and philosophy of the human future*, ed. M. More and N. Vita-More, 449–450. Chichester: Wiley-Blackwell.

[2] Bostrom, N. 2008. Why I want to be a posthuman when I grow up. In *Medical enhancement and posthumanity*, ed. Bert Gordijin and Ruth Chadwick, 107–136. Dordrecht: Springer. Available online https://nickbostrom.com/posthuman.pdf (accessed October 7, 2019).

[3] Faggella, D. 2018. Programmatically generated everything (PGE). August 27. https://danfaggella.com/programmatically-generated-everything-pge/ (accessed October 7, 2019).

[4] Yampolskiy, R. V., and S. Ziesche. 2018. Preservation of personal identity – A survey of technological and philosophical scenarios. In *Death and anti-death, volume 16: Two hundred years after Frankenstein*, ed. C. Tandy, 345–374. Ann Arbor, MI: Ria University Press.

[5] Faggella, D. 2018. The transhuman transition – Lotus eaters vs world eaters. May 27. https://danfaggella.com/the-transhuman-transition-lotus-eaters-vs-world-eaters/ (accessed October 7, 2019).

[6] Sandberg, A., and N. Bostrom. 2008. *Whole brain emulation: A roadmap*. Technical Report #2008-3. Future of Humanity Institute, Oxford University. https://www.fhi.ox.ac.uk/brain-emulation-roadmap-report.pdf (accessed October 7, 2019).

[7] Koene, R. A. 2012. Embracing competitive balance: The case for substrate-independent minds and whole brain emulation. In *Singularity hypotheses*, ed. A. H. Eden, E. Steinhart, D. Pearce, and J. H. Moor, 241–267. Berlin and Heidelberg: Springer.

[8] Tegmark, M. 2017. Substrate independence. In *This idea is brilliant: Lost, overlooked, and underappreciated scientific concepts everyone should know*, ed. J. Brockman, 162–166. New York: HarperCollins.

[9] Yampolskiy, R. V. 2019. Personal universes: A solution to the multi-agent value alignment problem. arXiv preprint arXiv:1901.01851. https://arxiv.org/pdf/1901.01851.pdf (accesssed October 7, 2019).

[10] Bostrom, N., A. Dafoe, and C. Flynn. 2018. Public policy and superintelligent AI: A vector field approach. Version 4.3. https://pdfs.semanticscholar.org/9601/74bf6c840bc036ca7c621e9cda20634a51ff.pdf (accessed October 7, 2019).

[11] Bostrom, N. 2018. The interests of digital minds. Draft 1.0. https://nickbostrom.com/papers/interests-of-digital-minds.pdf (accessed October 7, 2019).

[12] Tomasik, B. 2019. What are suffering subroutines? Updated May 17, 2019. https://reducing-suffering.org/what-are-suffering-subroutines/ (accessed October 7, 2019).

[13] Tomasik, B. 2019. Do video-game characters matter morally? Updated June 14, 2019. https://reducing-suffering.org/do-video-game-characters-matter-morally/ (accessed October 7, 2019).

[14] Ziesche, S., and R. V. Yampolskiy. 2019. Towards AI welfare science and policies. *Big Data and Cognitive Computing* 3(1)(Article #1). https://www.mdpi.com/2504-2289/3/1/2/htm (accessed October 7, 2019).

[15] Sotala, K., and L. Gloor. 2017. Superintelligence as a cause or cure for risks of astronomical suffering. *Informatica*, 41(4): 389–400. https://www.informatica. si/index.php/informatica/article/viewFile/1877/1098 (accessed October 15, 2019).

[16] Sloman, A. 1984. The structure of the space of possible minds. In *The mind and the machine: Philosophical aspects of Artificial Intelligence*, ed. S. B. Torrance, 73–82, Chichester: Ellis Horwood; New York: Halsted Press.

[17] Yudkowsky, E. 2008. Artificial intelligence as a positive and negative factor in global risk. In *Global catastrophic risks*, ed. N. Bostrom and M. M. Ćirković, 308–345. New York and Oxford: Oxford University Press.

[18] Yampolskiy, R. V. 2015. The space of possible mind designs. In *International Conference on Artificial General Intelligence*, 218–227. Cham: Springer.

[19] Warpefelt, H., and H. Verhagen. 2015. Towards an updated typology of non-player character roles. In *Proceedings of the International Conference on Game and Entertainment Technologies*, ed. K. Blashki and Y. Xiao, 131–139. International Association for the Development of the Information Society.

[20] Loosemore, R. 2014. Qualia surfing. In *Intelligence unbound: The future of uploaded and machine minds*, ed. R. Blackford and D. Broderick, 231–239. Chichester: Wiley-Blackwell.

[21] Bostrom, N. 2014. *Superintelligence: Paths, dangers, strategies*. Oxford: Oxford University Press.

[22] Bostrom, N., and E. Yudkowsky. 2014. The ethics of artificial intelligence. In *The Cambridge handbook of artificial intelligence*, ed. K. Frankish and W. M. Ramsey, 316–334. Cambridge: Cambridge University Press. https://nickbostrom.com/ ethics/artificial-intelligence.pdf (accessed October 7, 2019).

[23] Cowen, R., M. K. Stasiowska, H. Laycock, and C. Bantel. 2015. Assessing pain objectively: The use of physiological markers. *Anaesthesia* 70: 828–847.

[24] Skljarevski, V., and N. M. Ramadan. 2002. The nociceptive flexion reflex in humans – review article. *Pain* 96: 3–8.

[25] Chalmers, D. J. 1995. Facing up to the problem of consciousness. *Journal of Consciousness Studies* 2: 200–219. Available online: https://cogprints. org/316/1/consciousness.html (accessed October 7, 2019).

[26] Brooks, D. 2011. The limits of empathy. *New York Times*. September 29. https:// www.nytimes.com/2011/09/30/opinion/brooks-the-limits-of-empathy.html? scp=1&sq="thelimitsofempathy"&st=cse (accessed October 7, 2019).

[27] Pitrat, J. 2013. *Artificial beings: The conscience of a conscious machine*. London: ISTE Ltd; Hoboken, NJ: John Wiley & Sons.

[28] Yampolskiy, R. V. 2018. *Artificial Intelligence safety and security*. Boca Raton, FL: CRC Press.

[29] Pearce, D. 2007. The abolitionist project. https://www.abolitionist.com/ (accessed October 7, 2019).

[30] Bogomolov, A., B. Lepri, J. Staiano, N. Oliver, F. Pianesi, and A. Pentland. 2014. Once upon a crime: Towards crime prediction from demographics and mobile data. Orig. pub. in *Proceedings of the 16th International Conference on Multimodal Interaction*, 427–434. ACM. https://arxiv.org/pdf/1409.2983.pdf (accessed October 7, 2019).

3

Introducing the Concept of Ikigai to the Ethics of AI and of Human Enhancements

Soenke Ziesche and Roman V. Yampolskiy

3.1 Introduction to Ikigai and Time Use Research

Ikigai refers to a Japanese concept, which can be translated as "reason or purpose to live". It comprises those activities of life, which provide satisfaction and meaning.

Obviously, other cultures and philosophers have contemplated similar questions and concepts, such as the French "raison d'être" or the Greek "eudaimonia", yet for consistency, we use the term ikigai here, which captures what we intend to convey.

At the same time, in recent years, Western authors have "discovered" the topic of ikigai, which resulted in various popular scientific publications, [e.g., 1], and which is not further elaborated here since these approaches are often based on misconceptions and limited, e.g., by linking ikigai to somebody's professional career only. Therefore, the original works about ikigai are more appropriate as well as sufficient for us.

As Kamiya, one of the earliest ikigai theorists, pointed out, ikigai has various connotations [2]. An important distinction is described by the two aspects: ikigai kan and ikigai taishō, [e.g., 3]. Ikigai kan expresses feelings of satisfaction, wellbeing and a life worth living, and thus, it is a state of mind, while ikigai taishō comprises activities, experiences and circumstances, which lead to such feelings, and thus, it is rather a process. An important feature in Japanese culture is for individuals to reflect extensively towards a suitable ikigai taishō. Once a person has identified a personal ikigai taishō, she or he embraces the associated activities towards his or her subjective ikigai kan.

Examples for ikigai taishō include family life, hobbies, professional occupations and other social activities, especially, but not only such activities, which help others. Ikigai taishō can comprise a variety of activities from small everyday rituals to the pursuit of complex goals.

As we will specify below, one of the reasons for the interest in the concept of ikigai in recent years is that studies have revealed a correlation between

DOI: 10.1201/9781003565659-4

pursuing an individual ikigai and health and hence longevity. In other words, if humans have nothing what makes their life worth living, then this leads not only to unhappiness but also impacts their health and thus their life expectancy. Therefore, it is desirable to maximise the number of humans (or sentient digital minds in general, see below) who have found an ikigai and to allow them to pursue it.

In this regard, we also have to explore likely disruptions by upcoming technologies such as AI and human enhancements, e.g., through extended reality (XR). XR refers to any combinations of real and virtual environments, which are increasingly refined [4]. The topic of ikigai has so far not been linked to these technologies. Below we introduce visionary contemplations of XR, which look at the wellbeing and happiness of humans. Yet this is the static ikigai kan part, while the necessary ikigai taishō part (towards ikigai kan) appears to have been neglected, which would be the related day-to-day activities of humans in times of AI and human enhancements.

Due to both the significance of finding and pursuing an ikigai for humans as well as the expected disruption of our daily lives by AI and XR, it is argued here that the facilitation of ikigai should be an indispensable part of ethics for AI and XR. Two dimensions are foreseeable:

Currently, common ikigai activities may disappear, e.g., certain professional occupations, which are by many treated as their ikigai.

New, currently unknown ikigai activities may appear owing to new possibilities through AI and XR.

If this challenge is not prudently and timely addressed, we foresee what we call "risk of ikigai loss" or "i-risk". This can be seen as another level of the terms x-risks and s-risks introduced by others before. Bostrom coined and described x-risks or existential risks, which may lead to annihilation of life on earth [5]. Althaus and Gloor added the category of s-risks or suffering risks, which may increase suffering drastically [6]. While x-risks and s-risks are critical criteria to define undesirable scenarios, we argue that in addition, scenarios are to be avoided, in which humans are alive and not suffering per se but are devoid of any ikigai. This could be a world where ikigai activities have vanished or humans are prevented to find and pursue their ikigai. Risks, which may lead to such scenarios, we call i-risks.

In order to address i-risks, this chapter aims to introduce relevant synergies between the traditional concept of ikigai and the fields AI safety, human enhancement and AI welfare. For a scientific foundation, we introduce first the two concepts of metrics for ikigai and time use research.

3.1.1 Metrics for Ikigai

It can be objected that ikigai is an intangible concept; hence in order to scientifically discuss ikigai, metrics are required. We look at two dimensions of measurements:

To verify that having found and pursuing an ikigai is correlated with health and longevity.

To examine whether a suitable ikigai taishō for individual people can be calculated.

For the first topic, research has been conducted not only regarding ikigai in Japan but also in general regarding the purpose of life, [e.g., 7]. More specifically, e.g., Sone et al., Tanno et al. and Tomioka et al. concluded that, for Japanese people, having an ikigai is associated with health and longevity [8–10]. Since ikigai can have diverse manifestations, lists have been developed to assess multidimensional aspects of ikigai instead of merely asking subjects whether they found an ikigai. For example, Imai proposed "ikigai-9", which comprises nine sub-themes of ikigai [11].

For the second topic, no existing specific research has been found. However, e.g., Helliwell et al. analyse happiness and wellbeing in the annual "World Happiness Report". They also look at big data approaches and came to the conclusion "that Big Data is increasing the ability of researchers, governments, companies, and other entities to measure and predict the wellbeing and the inner life of individuals" (p. 119) [12].

While this is more associated with the static ikigai kan, we propose that big data could also be used to calculate related ikigai activities, i.e., ikigai taishō.[1] There are different categories of big data. For happiness measurement, especially user-generated digital content, e.g., on social media, is relevant, from which wellbeing can often be directly derived. Other big data may require more, usually AI supported, analysis to uncover information about individuals' tastes and choices. Overall, this may also lead to insights about the individuals' pursuit of ikigai activities. Also, a comparison with dating apps and the associated algorithms may be suitable. Algorithms could be developed and, through machine learning, improved, which suggest potentially matching ikigai activities to individuals instead of potentially matching partners.

We also note that big data carry significant privacy concerns in general and for ikigai research in particular, since the quest for ikigai is traditionally considered a private reflection. However, the benefits and drawbacks have to be weighed: If metrics and methods can be developed to find ikigai for people, thus, to increase their happiness and health, this may outweigh the privacy issues.

3.1.2 Time Use Research

Another relevant field of study for this chapter is time use research, which examines how humans on average allocate their time to certain activities. The main methodology in this regard is statistical time use survey, which are being conducted periodically in different countries. For example, Charmes provides a comparative analysis of 102 time use surveys carried out in 65 countries, which reveals significant differences of time use between different countries as well as between different sexes [14].

In order to allow comparison between time use surveys, standardised categories of activities are required. One leading example is the International Classification of Activities for Time-Use Statistics (ICATUS) from 2016, which has the following nine categories on the top level and various further sub-categories:

- Employment and related activities
- Production of goods for own final use
- Unpaid domestic services for household and family members
- Unpaid caregiving services for household and family members
- Unpaid volunteer, trainee and other unpaid work
- Learning
- Socialising and communication, community participation and religious practice
- Culture, leisure, mass media and sports practices
- Self-care and maintenance

The categories are mostly self-explanatory; hence, we just mention that sleeping is part of self-care and maintenance. An older but also relevant classification is the one by Ås, which distinguishes the following four categories [15]:

- Necessary time: Time spent on activities for physiological needs.
- Contracted time: Time that human beings spend to fulfil the contracts that they have made, most prominently for paid work.
- Committed time: Time committed to fulfil responsibilities, most prominently towards family members and possessions.
- Free time: The residual time left after performing contracted, committed and necessary time.

The last relevant classification introduced here is between primary and secondary time, which allows examining multitasking. In such situations, primary time refers to the main activity and secondary time to the side activity.

For our purposes, it is relevant to investigate 1) under which time use categories ikigai activities fall, 2) how much time people can devote to their individually ikigai activities and 3) how this may change in the future due to developments in AI and XR? As noted before, in particular, the first topic depends on individual preferences as, e.g., for some people, their employment is their ikigai, while others pursue their ikigai during social or leisure activities. And for the second topic, again the differences between countries, in particular between developing and developed countries, have to be highlighted. People may have an ikigai but may have hardly any time to pursue it, one out of many examples being women and girls in developing countries, who spend significant amounts of time on collecting water every day.

Time use research appears to have so far not been linked to ikigai. Gershuny examines relations between time use and wellbeing without referring to the distinction between ikigai kan and ikigai taishō [16]. For example, Gershuny compares countries and sex enjoyment ratings on a 0–10 scale for certain activities [16].

The introduced metrics for ikigai as well as time use research are revisited below when we examine how it could be avoided that human enhancement or AI lead to an increase of i-risks.

3.2 Ikigai and XR/Human Enhancement

3.2.1 Introduction

We define enhanced humans here as humans who still have biological parts, but whose capacities regarding health span, emotion and cognition (vastly) exceed current humans, e.g., by means of XR technologies or brain–computer interfaces. The field is progressing, and applications have, e.g., focused on entertainment [17], education [18], healthcare [19], manufacturing [20] and ethical self-assessment [21].

Usually, enhancements are described as beneficial for humans in various aspects and thus desirable, [e.g., 22]. Especially the potential achievement of enduring bliss is often mentioned and motivates research in the field, [e.g., 23]. Yampolskiy describes a particular scenario, which also addresses the AI value alignment problem (see further below), whereby personalised XR simulations for humans are created as Individual Simulated Universes tailored for individual values and happiness [24].

3.2.2 Challenge

As just introduced, the cited aim of XR or further enhancements is often for the concerned humans to obtain bliss as opposed to suffering, [e.g., 23]. Yet, here it is critical to reiterate the distinction between s-risks and i-risks: We argue that the absence of suffering is not a sufficient criterion for bliss, but bliss requires ikigai too, i.e., reason or purpose to live. In other words, scenarios with no or minimal suffering have no or low s-risks, but there may still be significant i-risks, which would concern non-suffering humans devoid of an ikigai. Non-suffering humans without ikigai exist also in our current world, yet we highlight that this situation may exacerbate as enhancement-related disruption may lead to the disappearance of traditional ikigai activities. Examples include the abolishment of professional occupations, which were considered as ikigai by some, such as teachers who could be replaced by XR-enabled education.

Regarding day-to-day activities in XR and enhancement scenarios, discussions have so far mostly focused on avoidance of boredom and availability of sufficient novelties for the concerned humans to experience [25–27]. Such deliberations about non-boring activities are relevant, especially for anticipated very long lifetimes owing to enhancements yet do not address exactly the issue of ikigai. In other words, an environment with manifold opportunities is helpful to find a purpose of life, but another step towards ikigai is required.

This challenge can be more specifically examined through time use statistics: since the space of possible activities for enhanced humans will be by definition very different from current day-to-day activities, their time use schedule will be modified. Table 3.1 illustrates how the relative time-share may change for the four categories: necessary time, contracted time, committed time and free time introduced by Ås [15]:

In addition to the relative shifts of time use, it is also likely that humans will overall have more time at hand for the following two reasons:

- Enhanced cognition could enable more efficient multitasking, i.e., formerly consecutive activities may be conducted in parallel during primary and secondary time.

- Enhancements are likely to increase the life span as well as health span.

While it will be up to future time use research to measure how exactly enhanced humans spend their time, for us, the focus is on the subset of ikigai taishō activities. Scenarios are imaginable in both directions that in a future world of XR and enhancements, the space of ikigai taishō activities will be restricted, e.g., through abolishment of professional occupations by enhancements, which were nevertheless considered as ikigai by some, as mentioned above, or, perhaps more likely, that the space of ikigai taishō activities will be enlarged, i.e., with the availability of more time as well as more options, especially in the category free time.

3.2.3 Ikigai-Related Desiderata

Ikigai research should be a critical component for framing ethical guidelines for content creators, developers, distributors and users of XR. The minimal

TABLE 3.1

Anticipated Time-Share Changes Due to XR and Enhancements

	Necessary Time	Contracted Time	Committed Time	Free Time
Examples while in enhanced/XR modus	None, because by definition necessary time is spent on physiological, i.e., non-enhanced/XR needs.	New professions enabled through enhancement/XR.	Responsibilities towards newly acquired possessions in XR/enhanced environment.	Entertainment and education in XR/enhanced environment.
Time-share enhanced/ XR modus compared with non- enhanced/XR modus	Same.	Smaller. Many professional occupations become obsolete.	Smaller. Many activities in this category will be taken over by AI systems and new technologies.	Larger. Takes over share of previously contracted and committed time-share.

goal should be that neither the available options nor the daily time for ikigai taishō activities would be shortened due to XR and further enhancements, while ideally XR and further enhancements provide for more options as well as even longer daily periods of ikigai taishō activities. However, the specifics of these ikigai taishō activities may be in current pre-enhancement times still unknown, yet as meaningful as well as satisfying.

Therefore, ethics for XR should carefully consider future time use from an ikigai-optimisation point of view, possibly integrated in context-sensitive frameworks such as Augmented Utilitarianism [13]. This should be in cooperation with time use researchers based on their established classifications, such as ICATUS, which may have to be extended. Ethics should especially consider the scenario that (large) parts of the known time use schedule might disappear; hence, the emerging vacuum has to be carefully filled. Recalling the current gap between inhabitants of developing and of developed countries concerning available time for pursuing ikigai activities, enhancements and the related ethics should in particular establish more equality in this regard. In fact, XR and enhancements may allow some humans for the first time to pursue ikigai activities as they may have been before entirely occupied by necessary and contracted time [15]. While activities during contracted time are actually considered by some as ikigai, the larger time share of free time means also more freedom in finding a personal ikigai.

In order to design and also to structure the unprecedented, largely artificial range of possibilities in XR and enhanced environments or even individual universes new occupations will be required. For example, Yampolskiy suggested in this regard the profession of a Universe Designer [28]. While preferably some or many traditional ikigai activities should continue to remain available in XR and enhancement environments and potentially for longer time periods per day, recalling the risk of boredom and depression over long lifetimes, further novel options should be explored too. Since the potential of ikigai activities for enhanced minds is both likely to be very different from what we know and impossible to forecast at this stage, an "Ikigai Designer" may be needed, similar to Yampolskiy's Universe Designer.

An advantage of such digital environments is that (big) data about activities, time use and states of minds from enhanced humans can be much easier retrieved, which may facilitate ikigai calculation, i.e., proven and tested ikigais from enhanced humans could be recommended to others.

In summary, the potential opportunities, thus desiderata, provided by XR and supported by dedicated ethics are as follows:

- Humans will likely find new formerly unknown ikigai taishō.
- Humans will likely have more freedom for ikigai activities because they may have less contracted and committed time but more free time.

- Humans will likely have more time for ikigai activities because they may conduct non-ikigai activities in a more time-efficient manner and they may live longer.

- In addition to more time and health, XR will likely furnish other resources, which are required for many traditional ikigai activities but tend to decline for ageing people [29], such as social networks.

3.3 Ikigai and AI Safety

3.3.1 Introduction

While a variety of problems towards AI safety has been identified, [e.g., 30], the focus here in connection with ethics for XR and further enhancements is on one of the most prominent ones, which is the value alignment problem. Yudkowsky and Bostrom described this problem and also pioneered this field [31, 32]. The basic question is how to cause an AI to pursue goals and values, which are aligned with human goals and values. A failure in solving this problem may constitute an existential threat to humanity, since there is no reason to assume that an AI will turn out to be value-aligned with humans without prior arrangements. The value alignment problem can be divided into sub-problems such as to agree on common human values, to precisely specify these values in a machine-understandable way and others. These sub-problems, thus the value alignment problem as a whole, have been proven to be very hard despite significant work in recent years, [e.g., 33, 34].

3.3.2 Challenge

Acknowledging that AI value alignment is indispensable as well as very hard to achieve, we argue that it is equally important that any AI must not impede the ikigai of individual humans because ikigai is vital for humans as shown above. To be more precise, we claim not only that ikigai is neglected by current AI value alignment research but also that achievement of AI value alignment would not necessarily entail that humans have and can pursue individual ikigai. Therefore, not taking ikigai into consideration for AI value alignment, research would constitute an i-risk, which we seek to avoid.

We illustrate the differences between values and ikigai in Table 3.2 as well as by an example scenario.

The following scenario provides an example when value alignment does not imply freedom of ikigai: it is assumed that there is an AI, which is aligned with human values but massively restricts the space of potential

TABLE 3.2

Comparison between Values and Ikigai

Values	Condition	Discrete	Common, Shared and Mutual	Should Be Aligned
Ikigai (taishō)	Activity	Holistic, big picture	Individual and diverse	Should be aligned (in the sense that they do not violate other people's values or impede other people's ikigai).

ikigai activities to, e.g., everyone playing musical instruments only (perhaps because the AI requires other matter as resources for its goals or the qualia of music creates pleasure for the AI). Playing musical instruments is indeed an ikigai for some people but not for all; thus, in this scenario, value alignment is achieved (as per assumption), while the pursuit of individual ikigai under the regime of such an AI is not accomplished. Therefore, lack of freedom of ikigai could also be taken as indicator for perverse instantiation issues [32].

Again, we stress that our focus is on day-to-day activities of humans in future times largely affected by powerful AIs. Since it has been shown that the pursuit of individual ikigai activities are linked to happiness and health and since the penetration of daily lives by AIs will have an impact of the space of possible activities, many humans would likely have to adjust or change their usual ikigai; hence, research on future ikigai activities is critical. Similar to XR and further enhancements, AIs are likely to influence the space of possible activities in both directions: ideally due to support by AIs, there may be less currently indispensable activities for humans to address their basic needs, including employment, which often constituted ikigai as well, but there could also be new so far unknown fulfilling activities.

3.3.3 Ikigai-Related Desiderata

Similar to the term "friendly AI" [31], an "ikigai-friendly AI" is desirable, of which three types are conceivable:

1. The AI does not prevent humans from searching, finding and pursuing their individual ikigai.

   Exceptional cases would be if a certain ikigai were in conflict with Omohundro's drives [35]. Examples are if the ikigai of an individual is to destroy AIs or if the ikigai of an individual and the AI both require the same resources.

2. In addition to the features of type 1, the AI has the knowledge how to help humans to find their individual ikigai and teaches it to individuals upon request.

A scenario would be that the AI gains the knowledge what would be fulfilling ikigai activities for individuals through big data analysis as introduced above. While, as mentioned, the quest for ikigai is traditionally considered a private reflection, support through an AI may be more sought-after, since the advent of powerful AIs may have disrupted and overwhelmed people's lives due to AI-related unemployment or other consequences.

This may include currently unknown ikigai activities, either because humans have never conceived them or these activities were only enabled by the AI, such as new kinds of hobbies or social interaction with artificial minds. These new ikigai activities could be more alien than innovative ikigais in XR environments due to the superiority of the AI and could be referred to as AI affordances [36], i.e., actions only made possible by an AI environment. This could be called AI effect, similar to the observer effect in physics. In other words, because of the AI, we may revise our whole concept of ikigai.

3. In addition to the features of types 1 and 2, the AI manipulates humans towards a specific ikigai. In that way, an AI environment may change our values and goals in an unforeseeable but positive manner. For example, an ikigai-friendly AI is aware of what is best for humanity, i.e., environmentalism, healthy lifestyle and nutrition, inclusion, overall as well as gender equality, anti-racism and anti-speciesism, and may then come up with supporting ikigai activities in this regard and indoctrinate humans to consider a convergent ikigai, which they would not have done otherwise.

   While this is a positive scenario of a manipulative AI—also negative, thus not ikigai-friendly—options are conceivable, which is that the AI manipulates or applies wireheading to humans, [e.g., 37], so that the affected humans consider activities, which support the AI goals, but which may include seemingly dull undertakings, as ikigai. An example would be that the AI seduces humans to produce paperclips in case the AI is a paperclip maximiser [38]. The difference to similar enslavement scenarios would be that the affected humans would not suffer but be under the impression to pursue their reason for being.[2]

An additional feature of all three types of "ikigai-friendly AI" could be that it recognises non-ikigai activities within individual time use schedules of humans and looks at ways how to take them over in order for the human to free time for her or his ikigai activities.

In summary, the potential opportunities provided by an ikigai-friendly AI are the same as mentioned above for XR, which includes potentially even more novel ikigai activities owing to AI affordances as well as more time for ikigai activities. However, there is also the risk of an "ikigai-unfriendly AI", which may not allow humans to pursue ikigai or which may wirehead humans in order to treat activities as ikigai that benefit the AI's goals but would usually not be entertained by humans.

3.4 Ikigai and AI Welfare

3.4.1 Introduction

It has been claimed that there is a non-zero probability that sentient digital beings not only exist but also that they may, at least temporarily, suffer, which may include AIs, [e.g., 40–42]. Therefore, Bostrom et al. developed a policy desideratum concerning the interests of sentient digital minds, which states, "AI is governed in such a way that maltreatment of sentient digital minds is avoided or minimized" (p. 18) [41]. In follow-up to this desideratum, Ziesche and Yampolskiy proposed a new field "AI welfare science" and offered recommendations for necessary activities towards the wellbeing of sentient digital minds as well as comprehensive anti-speciesism [42]. Certain not yet existing but in the future theoretically possible AIs have been categorised by Aliman et al. as Type II systems, "systems with a scientifically plausible ability to act independently, intentionally, deliberately and consciously and to craft explanations" (p. 2) (while Type I systems are defined as the complement of Type II systems with all present-day AIs being of Type I) [43].

3.4.2 Challenge

We distinguish two potential ikigai-related challenges for sentient digital beings:

1. If there is a non-zero probability that sentient digital beings exist, it could be argued that there is also a non-zero probability that sentient digital beings not only have the concept of ikigai but also the longing to pursue an ikigai. One of the defining features of sentient digital beings is the assumed capacity to perceive qualia. Ziesche and Yampolskiy look in their introductory paper mostly at suffering through perception of qualia of pain [42]. However, also other concerns could impact the wellbeing of humans as well as potentially the wellbeing of sentient digital minds. Bostrom et al. note as examples for other types of conceivable mistreatment of sentient digital minds restrictions to their autonomy, creativity and self-expression [41].

These restrictions can also be interpreted as lack of freedom or lack of opportunities for sentient digital minds to pursue an individual ikigai. In other words, concerned sentient digital minds without ikigai may suffer even if they do not perceive qualia of pain. This leads to a second dimension of the i-risks introduced above. On the one hand, if not carefully handled, many humans may lose their ikigai in times of AI and XR, as described above. On the other hand, AI and XR may produce (a high number of) sentient digital minds, which have the concept as well as the longing for ikigai, but may be prevented from finding and pursuing it. In other words, AI and XR may constitute i-risks not only because they may cause the number of humans without ikigai to increase but also because they may cause the number of other ikigai-requiring digital minds to increase.

2. As introduced, some ikigais may involve social activities, while others do not. We look here at the first category, the significance of which, e.g., Fukuzawa et al. highlighted by showing the relevance of social networks for ikigai [29].

XR technologies provide opportunities in this regard but also largely neglected risks. Not only many games but also other XR environments are based on interacting with (avatars of) other people. However, an integral part of social activities in XR are, apart from avatars of other humans, non-player characters (NPCs). And if indeed XR environments prove to provide innovative ikigais based on social activities, a large number of NPCs may be involved. Since NPCs could be sentient too [44] and hence could have a moral status, Ziesche and Yampolskiy argue that policies are required for NPCs towards ethical human enhancement and anti-speciesism [45]. Therefore, even if the above claim regarding i-risks for sentient digital minds is not true, XR environments, which support ikigais based on social activities, may increase suffering, i.e., pose an s-risk [6].

3.4.3 Ikigai-Related Desiderata

Ziesche and Yampolskiy developed recommendations for AI welfare policies regarding suffering of sentient digital minds, yet, as mentioned, initially limited to suffering through qualia of pain [42]. These should also be applicable for XR and AI environments, which have the purpose for humans to pursue their ikigai. Moreover, we propose to complement these recommendations as follows to also address suffering of sentient digital minds because of lack of ikigai:

- Initiate research on AI welfare science to develop methods to measure if sentient digital minds long for an ikigai.

- Potentially initiate research on AI welfare science to develop methods how individual ikigais could be found for sentient digital minds and how the pursuit of these individual ikigais by the sentient digital minds could be ensured.

Ziesche and Yampolskiy outline preliminary considerations and challenges regarding methodologies such as self-reporting as well as observation through functional and behavioural parameters [42].

Moreover, Ziesche and Yampolskiy proposed that humans are deterred from harming minds in other substrates by mandatorily equipping the humans with sensory perception of potential pain of the NPCs in order for the humans to understand when to stop activities because they cause the suffering [45]. Similarly, it could be made compulsory that humans in AI and XR environments are notified of frustration of NPCs prevented from finding and pursuing an ikigai, especially if the humans are engaged in their own individual ikigais based on social activities with these NPCs.

Future time use research for sentient digital minds may be useful in order to establish if sentient digital minds have time at all to allocate to ikigai activities. In this regard, one feature to take into account is that sentient digital minds may have a very different subjective rate of time [46].

Two points have to be highlighted:

- This section is largely speculation. Nevertheless, such prolegomena are relevant to reduce potential (massive) s-risks as well as i-risks in AI and XR environments, as initiated by Bostrom et al. [41] and as further specified by Ziesche and Yampolskiy [42].
- Anthropomorphic bias must be taken into account: 1) For other minds, ikigai may not be relevant, 2) they may have ikigais, which appear very alien to us and may be beyond our imagination, or 3) they may have obtained ikigai through wireheading, which may be sufficient and satisfying for them.

3.5 Summary and Future Work

In this chapter, we have argued that it is essential to add the topic of ikigai to the fields of ethical AI and human enhancement and to raise awareness for content creators, developers, distributors, and users of XR. In particular, we introduced the following overlooked challenges as well as suitable concepts:

i-risks concern scenarios, in which human and potentially other minds are not able to pursue their individual ikigai, and constitute a distinct level apart from x-risks and s-risks. We have identified unforeseen challenges for human enhancement, AI safety and AI welfare, which may lead to i-risks, as well as desiderata in this regard and some initial proposals for solutions.

Aspects that increase i-risks: XR and AI will likely be disruptive in a way that certain traditional ikigais may vanish, especially professional occupations. Also, AIs may restrict the freedom of humans to pursue their ikigai.

Aspects that reduce i-risks: If handled rightly, XR and (ikigai-friendly) AI may also provide affordances, opportunities and resources for new ikigais. Moreover, XR and AI could ensure more time for ikigai activities, including for those whose time use schedule hardly had slots for it before, i.e., XR and AI could reduce inequality among humans. To further reduce i-risks, also potential ikigai needs of sentient digital minds have to be taken into account.

We also proposed methods to measure and find ikigai for individual people through big data as well as extended time use research towards both XR and AI environments.

In this chapter, we have provided groundwork, but significant future work can be foreseen.

For the outlined desiderata, detailed specifications and solutions are required.

Moreover, the topic of ikigai and ethics could be widened to look at overall mental health policies in preparation for expected massive disruptions due to enhancements and AI. Since these disruptions may cause economic pressures as well as rapid social change and since the World Health Organization noted[3] that this may harm mental health, a contingency plan and related policies for a potential increase of mental health issues appear to be prudent to set up.

Notes

1 See also the recently proposed scheme of Augmented Utilitarianism in XR contexts [13].
2 For similar thoughts regarding AI-based religions, see also [39].
3 See https://www.who.int/news-room/fact-sheets/detail/mental-health-strengthening-our-response.

References

[1] García, H. & Miralles, F. Ikigai (2017). *Ikigai: The Japanese Secret to a Long and Happy Life*. Penguin.
[2] Kamiya, M. (1966). *Ikigai-ni-tsuite[On 755 ikigai]*. Tokyo, Japan: MisuzuShyobou.

[3] Mathews, G. (2008). Finding and keeping purpose in life. In G. Mathews & C. Izquierdo (Eds.), *Pursuits of happiness: Well-being in anthropological perspective* (pp. 167–186). Berghahn.

[4] Mann, S., Furness, T., Yuan, Y., Iorio, J., & Wang, Z. (2018). All reality: Virtual, augmented, mixed (x), mediated (x, y), and multimediated reality. *arXiv preprint arXiv:1804.08386*. https://arxiv.org/pdf/1804.08386.pdf (accessed 19 Sep 2020).

[5] Bostrom, N. (2002). Existential risks: Analyzing human extinction scenarios and related hazards. *Journal of Evolution and technology, 9*. https://nickbostrom.com/existential/risks.html (accessed 19 Sep 2020).

[6] Althaus, D., & Gloor, L. (2016). Reducing risks of astronomical suffering: A neglected priority. *Foundational Research Institute.* https://foundational-research.org/reducing-risks-ofastronomical-suffering-a-neglected-priority/ (accessed 19 Sep 2020).

[7] Crumbaugh, J. C., & Maholick, L. T. (1964). An experimental study in existentialism: The psychometric approach to Frankl's concept of noogenic neurosis. *Journal of Clinical Psychology, 20*(2), 200–207. https://citeseerx.ist.psu.edu/viewdoc/download?doi=10.1.1.505.6866&rep=rep1&type=pdf (accessed 19 Sep 2020).

[8] Sone, T., Nakaya, N., Ohmori, K., Shimazu, T., Higashiguchi, M., Kakizaki, M., … Tsuji, I. (2008). Sense of life worth living (ikigai) and mortality in Japan: Ohsaki Study. *Psychosomatic Medicine, 70*(6), 709–715. https://se-realiser.com/wp-content/uploads/2018/11/Sone-et-al.-Ikigai-and-Mortality-in-Japan-2008.pdf (accessed 19 Sep 2020).

[9] Tanno, K., Sakata, K., Ohsawa, M., Onoda, T., Itai, K., Yaegashi, Y., … JACC Study Group. (2009). Associations of ikigai as a positive psychological factor with all-cause mortality and cause-specific mortality among middle-aged and elderly Japanese people: findings from the Japan Collaborative Cohort Study. *Journal of Psychosomatic Research, 67*(1), 67–75.

[10] Tomioka, K., Kurumatani, N., & Hosoi, H. (2016). Relationship of having hobbies and a purpose in life with mortality, activities of daily living, and instrumental activities of daily living among community-dwelling elderly adults. *Journal of Epidemiology, 26*(7), 361–370. https://www.jstage.jst.go.jp/article/jea/26/7/26_JE20150153/_pdf (accessed 19 Sep 2020).

[11] Imai, T. (2012). The reliability and validity of a new scale for measuring the concept of Ikigai (Ikigai-9). *[Nihon koshu eisei zasshi] Japanese Journal of Public Health, 59*(7), 433–439.

[12] Helliwell, J. F., Layard, R., & Sachs, J. D. (2019). *World happiness report 2019.* https://s3.amazonaws.com/happiness-report/2019/WHR19.pdf (accessed 19 Sep 2020).

[13] Aliman, N.-M., Kester, L. & Werkhoven, P. J. (2019). XR for Augmented Utilitarianism. In *IEEE International Conference on Artificial Intelligence and Virtual Reality (AIVR)*, 283–285.

[14] Charmes, J. (2015). Time use across the world: Findings of a world compilation of time use surveys. *UNDP Human Development Report Office, background Paper, New York.* https://www.hdr.undp.org/sites/default/files/charmes_hdr_2015_final.pdf (accessed 19 Sep 2020).

[15] Ås, D. (1978). Studies of time-use: Problems and prospects. *Acta Sociologica, 21*(2), 125–141. https://www.timeuse.org/sites/default/files/public/ctur_journal_article/2179/DagfinnAs.pdf (accessed 19 Sep 2020).

[16] Gershuny, J. (2011). Time-use surveys and the measurement of national well-being. *Centre for Time Use Research, University of Oxford, Swansea, UK, Office for National Statistics.* https://www.timeuse.org/sites/ctur/files/public/ctur_report/4486/timeusesurveysandwellbein_tcm77-232153.pdf (accessed 19 Sep 2020).

[17] Von Itzstein, G. S., Billinghurst, M., Smith, R. T., & Thomas, B. H. (2019). *Augmented Reality Entertainment: Taking Gaming Out of the Box.* https://www.researchgate.net/profile/Bruce_Thomas3/publication/318183569_Augmented_Reality_Entertainment_Taking_Gaming_Out_of_the_Box/links/5e92eddf4585150839d65361/Augmented-Reality-Entertainment-Taking-Gaming-Out-of-the-Box.pdf (accessed 24 Oct 2020).

[18] Liu, D., Dede, C., Huang, R., & Richards, J. (Eds.). (2017). *Virtual, augmented, and mixed realities in education.* Singapore: Springer.

[19] Andrews, C., Southworth, M. K., Silva, J. N., & Silva, J. R. (2019). Extended reality in medical practice. *Current Treatment Options in Cardiovascular Medicine, 21*(4), 18. https://www.ncbi.nlm.nih.gov/pmc/articles/PMC6919549/ (accessed 24 Oct 2020).

[20] Fast-Berglund, Å., Gong, L., & Li, D. (2018). Testing and validating Extended Reality (xR) technologies in manufacturing. *Procedia Manufacturing, 25,* 31–38. https://pdfs.semanticscholar.org/316e/03ddb48e56dfd73ce4cbc6a72e059bb2d46e.pdf (accessed 24 Oct 2020).

[21] Aliman, N.-M. & Kester, L. (2019). Extending socio-technological reality for ethics in artificial intelligent systems. In *IEEE International Conference on Artificial Intelligence and Virtual Reality (AIVR),* 275–282.

[22] Bostrom, N. (2008). Why I want to be a posthuman when I grow up. In *Medical enhancement and posthumanity.* Springer, Dordrecht: 107–136. https://nickbostrom.com/posthuman.pdf (accessed 19 Sep 2020).

[23] Faggella, D. 2019. The Transhuman Transition – What it is and Why it Matters. https://danfaggella.com/transhuman-transition/ (accessed 19 Sep 2020).

[24] Yampolskiy, R. V. (2019). Personal universes: A solution to the multi-agent value alignment problem. arXiv preprint arXiv:1901.01851. https://arxiv.org/pdf/1901.01851.pdf (accessed 19 Sep 2020).

[25] Ziesche, S., & Yampolskiy, R. V. (2016). *Artificial Fun: Mapping Minds to the Space of Fun.* arXiv preprint arXiv:1606.07092. https://arxiv.org/ftp/arxiv/papers/1606/1606.07092.pdf (accessed 19 Sep 2020).

[26] Loosemore, R.P.W. (2014). Qualia Surfing. In R. Blackford & D. Broderick (Eds.), *Intelligence unbound: The future of uploaded and machine minds.* Chichester: John Wiley & Sons, 231–239.

[27] Yudkowsky, E. (2009). *The Fun Theory Sequence.* https://lesswrong.com/lw/xy/the_fun_theory_sequence/ (accessed 19 Sep 2020).

[28] Yampolskiy, R.V. (2017). Future Jobs – The Universe Designer. *Circus Street.* https://www.circusstreet.com/blog/future-jobs-the-universe-designer (accessed 19 Sep 2020).

[29] Fukuzawa, A., Katagiri, K., Harada, K., Masumoto, K., Chogahara, M., Kondo, N., & Okada, S. (2019). A longitudinal study of the moderating effects of social capital on the relationships between changes in human capital and ikigai among Japanese older adults. *Asian Journal of Social Psychology, 22*(2), 172–182. https://onlinelibrary.wiley.com/doi/pdf/10.1111/ajsp.12353 (accessed 19 Sep 2020).

[30] Amodei, D., Olah, C., Steinhardt, J., Christiano, P., Schulman, J., & Mané, D. (2016). Concrete problems in AI safety. arXiv preprint arXiv:1606.06565. https://arxiv.org/pdf/1606.06565.pdf%20https://arxiv.org/abs/1606.06565.pdf (accessed 19 Sep 2020).

[31] Yudkowsky, E. (2008). Artificial intelligence as a positive and negative factor in global risk. In N. Bostrom & M.M. Ćirković (Eds.), *Global Catastrophic Risks*. New York: Oxford University Press, 308–345. https://intelligence.org/files/AIPosNegFactor.pdf (accessed 19 Sep 2020).

[32] Bostrom, N. (2014). *Superintelligence: Paths, dangers, strategies*. Oxford, UK: Oxford University Press.

[33] Soares, N. (2015). The value learning problem. Machine Intelligence Research Institute, Berkeley. https://citeseerx.ist.psu.edu/viewdoc/download?doi=10.1.1.674.6424&rep=rep1&type=pdf (accessed 19 Sep 2020).

[34] Turchin, A., & Denkenberger, D. (2019). Literature Review: What Artificial General Intelligence Safety Researchers Have Written About the Nature of Human Values. https://philpapers.org/rec/TURLRW (accessed 19 Sep 2020).

[35] Omohundro, S.M. (2007). The nature of self-improving artificial intelligence. Singularity Summit. https://selfawaresystems.files.wordpress.com/2008/01/nature_of_self_improving_ai.pdf (accessed 19 Sep 2020).

[36] Nye, B. D. & Silverman, B. G. (2012). Affordances in AI. In N. M. Seel (Ed.), *Encyclopedia of the Sciences of Learning*. New York, NY: Springer, 183–187. https://repository.upenn.edu/cgi/viewcontent.cgi?article=1677&context=ese_papers (accessed 19 Sep 2020).

[37] Yampolskiy, R. V. (2014). Utility function security in artificially intelligent agents. *Journal of Experimental & Theoretical Artificial Intelligence, 26*(3), 373–389. https://cecs.louisville.edu/ry/Utility.pdf (accessed 19 Sep 2020).

[38] Bostrom, N. (2003). Ethical issues in advanced artificial intelligence. In *Science fiction and philosophy: from time travel to superintelligence*, 277–284. https://www.fhi.ox.ac.uk/wp-content/uploads/ethical-issues-in-advanced-ai.pdf (accessed 19 Sep 2020).

[39] Ziesche, S. (2019). An AI May Establish A Religion. In *Death and anti-death*, ed. C. Tandy, Volume 17: 309–334. Ann Arbor: Ria University Press.

[40] Tomasik, B. (2011). *Risks of astronomical future suffering*. Foundational Research Institute: Berlin, Germany. https://longtermrisk.org/files/risks-of-astronomical-future-suffering.pdf (accessed 19 Sep 2020).

[41] Bostrom, N., Dafoe, A., & Flynn, C. (2018). Public Policy and Superintelligent AI: A Vector Field Approach. *Governance of AI Program, Future of Humanity Institute, University of Oxford: Oxford, UK*. https://nickbostrom.com/papers/aipolicy.pdf (accessed 19 Sep 2020).

[42] Ziesche, S. & Yampolskiy, R. V. (2019a). Towards AI Welfare Science and Policies. *Big Data and Cognitive Computing, 3*(1), 2. https://www.mdpi.com/2504-2289/3/1/2/htm (accessed 19 Sep 2020).

[43] Aliman, N. M., Elands, P., Hürst, W., Kester, L., Thórisson, K. R., Werkhoven, P., Yampolskiy, R. & Ziesche, S. (2020). Error-Correction for AI Safety. In *International Conference on Artificial General Intelligence*, 12–22. Cham: Springer.

[44] Tomasik, B. (2014). Do video-game characters matter morally? Updated June 14, 2019. https://reducing-suffering.org/do-video-game-characters-matter-morally/ (accessed 19 Sep 2020).

[45] Ziesche, S., & Yampolskiy, R. V. (2019b). Do No Harm Policy for Minds in Other Substrates. *Journal of Evolution & Technology*, *29*(2). https://jetpress.org/v29.2/ziesche.html (accessed 19 Sep 2020).

[46] Bostrom, N., & Yudkowsky, E. (2014). The ethics of artificial intelligence. *The Cambridge Handbook of Artificial Intelligence*, *1*, 316–334. https://faculty.smcm.edu/acjamieson/s13/artificialintelligence.pdf (accessed 19 Sep 2020).

4

Mapping the Potential AI-Driven Virtual Hyper-Personalised Ikigai Universe

Soenke Ziesche and Roman V. Yampolskiy

> Before that, he'd written over three hundred comic operas, with librettos in Italian, French and English—and staged most of them, with puppet performers and audience. Before that, he'd patiently studied the structure and biochemistry of the human brain for sixty-seven years.
>
> *Greg Egan: Permutation City*

4.1 Introduction

The Japanese concept of ikigai can be translated as "reason or purpose to live". It comprises those activities of life, which give humans satisfaction and meaning. The concept of ikigai has been popularised in Western contexts in recent years, yet often misinterpreted, as has been stressed [1, 2] and as can be elucidated by reverting to primary Japanese literature [e.g., 3].

Ikigai has various connotations, two of which are referred to as ikigai kan and ikigai taishō [3]. Ikigai kan encompasses feelings of satisfaction, wellbeing and a life worth living, and thus, it is a state of mind, while ikigai taishō describes activities, experiences and situations, which create such feelings, and thus, it is rather a process.

It can be stated that it is very much desirable for humans to have found an ikigai. Therefore, scenarios, in which (a high number of) humans are devoid of any ikigai, ought to be prevented. Such scenarios have been coined "i-risk scenarios" [2] and constitute a distinct level of risks, supplemental to previously defined s-risk scenarios [4] and x-risk scenarios [5], which stand for "suffering risk" and "existential risk", respectively, i.e., scenarios where humans (severely and continuously) suffer or become extinct.

Developments in AI and other emerging technologies may lead to i-risk scenarios [2]. This comprises situations in which AI and other emerging technologies take over much more efficient activities, which humans used to carry out day by day and considered them as ikigai taishō activities. As

DOI: 10.1201/9781003565659-5

a result, the affected humans may struggle to find the reason for which to get up in the morning and how to meaningfully spend the hours of the day, which are other paraphrases for ikigai.

Therefore, the purpose of this chapter is to explore ways to reduce i-risks by applying AI systems for the creation of innovative virtual ikigai taishō activities, which could lead to ikigai kan. This endeavour is timely, since 1) i-risks are a serious concern, but 2) have largely been neglected so far and 3) AI systems together with virtual worlds have the potential to alleviate i-risks, especially given recent developments in these fields. While the third point is the main topic of this chapter, the first two points are briefly described below:

i-risks ought to be reduced, since several studies have shown that having an ikigai has a positive impact on health and wellbeing [for overviews: 6, 2]. Therefore, it is not desirable if the space of potential ikigais for humans is being reduced. In contrast, when it comes to developments in AI and other emerging technologies, it is not only largely overlooked that those may reduce the space of potential ikigais, but there are also no efforts made to harness AI and other emerging technologies for innovative virtual ikigai taishō activities. Instead, prime applications of virtual worlds, e.g., the meta-verse, that are discussed are simulations, games, office, social, marketing and education [7]. While some of these, e.g., social and education, may evolve into ikigais as described below, targeted undertakings towards ikigai taishō activities when creating virtual worlds have not been observed.

This chapter is structured as follows: First recent developments in AI and their links to i-risks are described. This is followed by the two main parts of the chapter, which outline the potential virtual ikigai universe as well as the potential hyper-personalised ikigai universe, supported by these developments in AI. The chapter concludes with a summary and with ideas for future investigation.

4.1.1 Generative AI

Since the issue of i-risks was raised for the first time [2], further significant developments in AI took place, of which advances in generative AI are most remarkable as well as most relevant for i-risks, in a positive as well as a negative way as will be outlined.

Generative AI is the umbrella term for machine learning algorithms, which generate artificial digital content such as text, images, audio and video content, based on large amounts of training data. The quality of this content is increasingly of an extent that humans cannot distinguish whether the content has been created by a machine or a human. The technique most of them use is called large language model or transformer. Details are omitted here as well as the various challenges these systems have, such as bias, stereotypes and creation of disinformation [e.g., 8].

The two points, which are relevant here, are 1) the assumption that these systems will continue to improve based on even more training data input as well as more powerful hardware as it was demonstrated in the recent AI history[1] and 2) that generative AI has a significant impact on i-risks.

4.1.2 Arising i-risks

Various aspects of the reduced space of ikigai against the backdrop of developments in AI and other emerging technologies have been outlined before [2]. Humans who have lost or will lose their ikigai due to latest developments in AI are mostly those whose professional occupations, which they have treated as their ikigai, have vanished or will vanish.[2]

While before this concerned rather monotonous routine tasks, which could be easily automated, it is likely to affect in the (near) future also creative tasks due to the advances in generative AI, as described above. Examples comprise writing (including novels, poems, computer code, movie scripts, instruction manuals and advertising texts), designing (including graphics, branding, animations, cooking recipes and fashion), composing, painting and photographing. In other words, many professions in these fields will likely become obsolete. Of course, it is possible for humans to continue with these activities during leisure time and not for monetary compensation, but it will be frustrating if much superior AI systems massively outperform them and produce oeuvres of much higher quality in much shorter time.

These advances in generative AI initially increase i-risks but, as will be outlined below, also provide the opportunity to reduce i-risks. The increase is caused by the expectation that many people will lose their jobs in creative sectors due to far superior generative AI; thus, these people will also lose their ikigai. There are no specific numbers, since this is an ongoing process. Potential proxy indicators could be suicide rate, drug addiction or the number of hikikomori, which is a form of social withdrawal, observed in Japan but also other countries [e.g., 9].

4.1.3 AI-Driven Hyper-Personalisation

The second field of AI advances, which is critical to reduce i-risks through virtual opportunities and thus is introduced here briefly, is AI-driven hyper-personalisation. This field can be divided into hyper-personalised content and hyper-personalised interaction.

The first group comprises tailored individual content for humans based on large amounts of data about the particular human, analysed by AI. Naturally, this is interesting for the commercial sector and has been applied in marketing [e.g., 10] and entertainment [e.g., 11]. In brief, pertinent data include geographics, demographics and psychographics, among others, which are traced through website analytics and consumer behaviour and which are used to generate personalised content, such as personalised messages, personalised websites and personalised product recommendations. AI-driven hyper-personalisation has not yet been used to identify suitable ikigai taishō activities for humans, except to some extent for the sub-field of learning [e.g., 12].

The field of learning can be also used to illustrate the second group introduced above, hyper-personalised interaction. AI-driven systems for hyper-personalised education analyse performance data of students in real-time and establish individual learning plans according to the strengths and weaknesses of the student, which replace the prevailing but in many aspects ineffective one-size-fits-all education system.

Other fields where AI-driven hyper-personalised interaction is applied are, e.g., health [e.g., 13] or marketing for social good [e.g., 14]. For AI-driven hyper-personalised interaction, also non-player characters (NPCs) or assistants are being used in various sectors such as education, wellbeing [for both e.g., 15] or marketing [e.g., 16].

As above, for generative AI, also AI-driven hyper-personalisation challenges related to privacy issues, ownership of data, bias [e.g., 12], as well as addiction to hyper-personalised content [for social networking e.g., 17] are not discussed here. The bottom line here is that AI-driven hyper-personalisation has not been used yet to identify individual ikigais.

In a more comprehensive theoretical overview, it has been described how the human needs such as novelty, romantic love and achievements are likely to be fulfilled by an increasingly enhanced immersive experience in precisely tailored manner through AI-driven hyper-personalisation in the future [18]. While in this overview it is not referred to ikigai, the used phrase "need for a sense of achievement" resembles the concept of ikigai in some aspects.

4.2 Virtual Ikigai Universe

As outlined, there is a risk that the space of ikigai taishō activities in the real world declines. This means that to counter i-risks urgently, new ikigais are required. Therefore, this chapter looks for upcoming virtual worlds as a new space for ikigais, especially virtual worlds, which are potentially created by generative AI, given the current promising developments in this field. Currently, the main application of virtual worlds is entertainment, with some exceptions. Yet, up to now, virtual worlds have hardly been considered for ikigai taishō activities. And as will be explained also below, entertainment must not to be confused with ikikai. Those often mindless and passive entertainment activities in virtual space may even lead to addiction, which is certainly not an ikigai.

Therefore, the desideratum is that AI systems create virtual worlds where ikigai taishō activities can be practised, including potentially new, thus formerly unknown ikigais.

4.2.1 Ikigai Taxonomy

Detailed taxonomies of ikigai taishō activities have not been found in the literature. A categorisation of the following five types of ikigai taishō has been provided [1]:

- Yarigai: Things worth doing
- Asobigai: The value of playing
- Hatarakigai: Work worth doing
- Manabigai: The value of learning
- Oshiegai: The value of teaching

While it appears desirable to refine this classification, it is sufficient for this chapter to explore how AI-driven virtual counterparts for each of these categories could look like.

4.2.2 AI-Driven Virtual Counterparts

This section serves as the centrepiece of this chapter. The aim is to map innovative virtual ikigai taishō day-to-day activities when not only several offline ikigai taishō activities have vanished due to developments in AI and other emerging technologies but also more time is available [2].

Our hypothesis is that for many if not most known ikigai taishō activities, there will be AI-driven virtual counterparts, which are likely to be complemented by yet unknown ikigai taishō activities, which may emerge owing to new possibilities through AI.

Another aspect is the unsolved question whether there are sentient digital minds or could be created in the future [e.g., 19]. This is relevant because ikigai taishō activities often involve social activities; thus, virtual counterparts require NPCs. While sentient NPCs would provide further opportunities for virtual ikigai taishō activities, they would also be moral patients and thus must not be harmed, intentionally or unintentionally [20].

Below potential AI-driven virtual counterparts of the five types of ikigai taishō as introduced above are presented.

4.2.2.1 Yarigai: Things Worth Doing

Yarigai comprises a wide range of things. Numerous small or bigger activities may constitute ikigai taishō for individual humans, ranging from a walk in the nature to caring for a pet.

As AI-driven virtual counterparts, it is conceivable that, instead of a walk in the real nature, immersive navigating in complex and completely unknown virtual worlds could be explored [e.g., 21].[3] This could also be a virtual replica of an existing place on earth and/or in another historical epoch or on another existing planet. The experiences could be massively augmented, i.e., beyond our normal senses. For example, during a virtual nature walk, we may be able to zoom in to inspect tiny insects, listen to the sounds of bats or look at a whole forest with a bird's eye view. Moreover, this could be hyper-personalised as outlined below to match the ikigai preferences of the individual human as exactly as possible.

As for caring for a pet, it is conceivable that in a virtual world, completely new creatures appear and need attention as well as care. Again, instead of creatures that do not exist on earth, it could be realistic replicas of cats or dogs, or even extinct animals. Moreover, this environment could be hyper-personalised, e.g., in a way that this creature appears to the individual human as cuter than anything s/he has seen before. As introduced above, in such a scenario, for now the unsolved question is relevant: whether NPCs, such as virtual pets, are sentient; and in which case they must be treated as a moral patient [20]. It must be noted that virtual pets, such as the Tamagotchi, exist already [22], but 1) they have not been considered for an ikigai mapping and 2) it is likely that they become much more sophisticated through AI-driven hyper-personalisation.

It would be understandable if this initially did not sound attractive to many humans, also given that currently the quality of virtual worlds is not very advanced. However, it has to be considered 1) that developments in AI are progressing fast, as indicated above, and 2) that in the past decades, humans adjusted to various technologies, which were unimaginable for previous generations [23]. This may sound outlandish, but in the end, there is no significant difference whether an ikigai is to collect real stamps or certain items in the virtual world.[4]

4.2.2.2 Asobigai: The Value of Playing

For this category, it is important to separate it from games and virtual entertainment, which are comprised by asobigai, but not topic of this chapter. Yet asobigai encompasses also social interactions [1], which are considered ikigai taishō activities by many and for which here AI-driven virtual counterparts are outlined [see also: 24]. Therefore, for this category, NPCs are critical, which may be human-like or other creatures. Important is that the human and the NPCs have a common language, noting that, in virtual worlds, also other creatures could speak human language or have other means to communicate. Complex interactions with a variety of NPCs can be imagined, which include talking about different topics, problems of the concerned human or philosophical questions but also gossiping and joking. As described below under hyper-personalisation, the NPCs will be perfectly adjusted in terms of knowledge and empathy.

4.2.2.3 Hatarakigai: Work Worth Doing

Hatarakigai refers to cases when the professional work of someone is considered by her or him as ikigai. As outlined above, i-risks for hatarakigai may be especially high, taking into account jobs that disappear due to AI and other emerging technologies.

However, AI-driven virtual counterparts for many professions are conceivable. Doctors could treat simulated humans, non-human animals or completely different creatures [25]. Researchers could study the physics, chemistry, biology, astronomy, sociology, history and so on of completely different worlds and their inhabitants. Creative humans could design anything in these worlds from fashion, machines, houses, landscapes to whole universes [26].

4.2.2.4 Manabigai: The Value of Learning

Another common ikigai taishō activity is to learn something, for which the real world also presents many opportunities.

In addition, a variety of AI-driven virtual counterparts can be imagined. For example, if the individual human likes to learn languages, the AI could create a completely new language for the human to learn, in written and verbal form. Also, the just mentioned sciences of completely different worlds and their inhabitants, such as physics, chemistry, biology, astronomy, sociology or history, provide for extensive learning content. The learning methodology may differ significantly from traditional ones and may include immersive experiences of virtual environments as well as AI-driven hyper-personalised individual learning plans according to the strengths and weaknesses of the student.

4.2.2.5 Oshiegai: The Value of Teaching

Likewise, teaching is considered by many as ikigai taishō activity, as part of their profession, but also in other contexts.

As for AI-driven virtual counterparts, NPCs would have to serve as students, and humans can teach them either knowledge from the real world or knowledge they gained in virtual worlds while embracing virtual manabigai.

This concludes a glimpse of possible AI-driven virtual counterparts for ikigai taishō activities, while it must be stressed that there are considerable unknown unknowns, given that AI systems are much smarter than humans and may come up with highly satisfying ikigai taishō activities, currently unimaginable for us.

4.3 Hyper-Personalised Ikigai Universe

After describing the possibility of innovative virtual ikigai taishō worlds, it
has to be examined how specific ikigai taishō activities can be matched with
individual humans, since ikigais are very personal as indicated. AI-driven
hyper-personalisation is key for this undertaking, and two sub-steps can be
distinguished:

1) Content: This involves AI systems calculating based on big data a
   suitable virtual ikigai for individual humans.
2) Interaction: When the human conducts ikigai taishō activities, the
   AI system provides feedback, which has two purposes: To adjust
   the ikigai taishō activities based on various parameters in order to
   further optimise the resulting ikigai kan feeling and to praise the
   human for the way s/he conducts the ikigai taishō activities.

Therefore, the desideratum is that AI not only identifies within the space of
virtual ikigai taishō worlds, as described above, for individuals their suitable hyper-personalised virtual ikigai taishō world but also provides hyper-
personalised feedback, while the individuals practise their ikigai taishō
activities.

4.3.1 Ikigai Metrics

The precondition for AI-driven hyper-personalisation of ikigai is the availability of ikigai metrics. However, not many such metrics exist [2], one of the
exceptions being ikigai-9. This approach consists of the following nine statements, which could be seen as sub-components of ikigai kan [27]:[5]

- I believe that I have some impact on someone.
- My life is mentally rich and fulfilled.
- I am interested in many things.
- I feel that I am contributing to someone or to society.
- I would like to develop myself.
- I often feel that I'm happy.
- I think that my existence is needed by something or someone.
- I would like to learn something new or start something.
- I have room in my mind.

It is desirable if there were more metrics, yet again big data and AI provide
for an opportunity: when pursuing virtual ikigai taishō activities, many

more data are recorded, which can be harnessed for ikigai calculation as well as optimisation. As it is a feature of machine learning algorithms to discover patterns and trends unbeknown to humans in data, there is a likelihood that AI may reveal further ikigai metrics, i.e., data and parameters, which measure to what extent activities in the virtual world contribute to ikigai kan feelings. Therefore, these activities constitute ikigai taishō activities for particular humans, keeping in mind that ikigai is very personal and differs significantly among humans.

4.3.2 AI-Driven Matching

As introduced above, this field can be divided into hyper-personalised content and hyper-personalised interaction.

4.3.2.1 Content

As also introduced above, generative AI will likely be capable to create a broad range of innovative potential virtual ikigais. These ikigai options have then to be matched with the preferences of individual humans as exactly as possible. The goal of this process would be the creation of a virtual environment with the specific purpose of being the ikigai taisho of a specific human.

This could be initiated with a survey about the individual's interests and likings. AI systems would then recommend proven and tested ikigais based on patterns learned from a large range of data—a process comparable with dating apps or suggestions to customers based on previous purchases.

4.3.2.2 Interaction

The proposed ikigai taisho activities, thus the customised virtual world, could then be improved and refined based on further data, which the AI system receives through the behaviour patterns of the human in this world. An example would be that an initial ikigai taisho activity was to explore virtually simulated parts of planet earth, which was then adjusted to the exploration of other planets and celestial bodies and later further extended to the exploration of fantasy worlds. This process can be continued until the individual has identified for him- or herself ideal virtual ikigai taishō activities, which lead to perfect bliss and feelings of satisfaction and thus ikigai kan. Therefore, AI-driven hyper-personalisation of ikigai taishō activities is an essential complement to the AI-driven creation of an ikigai-suitable virtual world, which does not exist in this sophisticated manner in the real world.

As mentioned above, the second element of interaction comprises accolades and social validation. While this differs from human to human, not everyone requires such feedback to feel bliss, and the ikigai-9 components above illustrate that six of them are not linked to social validation and the three of them are as follows:

- I believe that I have some impact on someone.
- I feel that I am contributing to someone or to society.
- I think that my existence is needed by something or someone.

In a hyper-personalised virtual world, the feedback to these three ikigai-9 components would be provided by NPCs, which are empathic as well as perfectly adjusted to the individual human and her or his ikigai. Since the AI will find out precisely the degree of social validation that satisfies the human, the feedback will be much more rewarding than in the real world by creating the thoughts and feelings the human is craving for.

4.4 Further Ideas and Summary

4.4.1 Further Ideas

This chapter aims to be hands-on for the reduction of i-risks in the near future, while at a later stage, the space of potential ikigais for humans may be further increased due to emerging technologies such as human enhancements, brain–machine interfaces and uploading as well as additional developments in AI (for an overview, see [2]). One particular progress of brain–machine interfaces would be the direct control of human qualia in a way that the pursuit of virtual hyper-personalised ikigai taishō activities could be directly rewarded with sensory pleasure.

As AI may approach Artificial General Intelligence, another AI risk has to be considered, in addition to the mentioned one that developments in AI may increase i-risks, which is that AI systems may turn out not be ikigai-friendly. It is beyond the scope of this chapter, but the values of a non-ikigai-friendly AI system would not be aligned with the values of humans. Thus, the goals of such an AI system and humans would be conflicting. Thus, such an AI system would not support humans in finding their individual ikigai or would even prevent them from pursuing their individual ikigai [2].

Another consideration is related to the prevailing global inequalities. One step to counteract this plight would be to provide virtual hyper-personalised ikigai as a digital public good for everybody. In other words, policies and legislation have to be developed and implemented so that everyone gets access to his or her virtual hyper-personalised ikigai universe [also: 29]. This may also be a remedy to reduce the time humans spent online for (potentially addictive) entertainment activities.

And yet an additional aspect is that this chapter may offer another explanation, if we live in a simulation [25], for a potential motivation of the simulator, which is that running this simulation may be the virtual ikigai of this being.

4.4.2 Summary

i-risks have an intersecting set with AI risks in the sense that certain developments in AI increase i-risks. Yet, it has been outlined that there are also AI-driven virtual opportunities, which reduce i-risks. This chapter introduces concrete measures to tackle i-risks by mapping the potential AI-driven virtual hyper-personalised ikigai universe. It has been illustrated that advances in generative AI, virtual worlds and AI-driven hyper-personalisation provide for opportunities to constitute spaces for formerly unknown ikigais.

It has to be emphasised that the focus here is on ikigai taishō activities (which ideally lead to ikigai kan feelings of satisfaction). This is critical as technologies are likely to free up large amounts of time for many humans. Therefore, it may become a challenge for humans to fill their day with activities and experiences that have a sense of purpose.

While, as mentioned, it has been shown that having an ikigai may have a positive impact on health and wellbeing, spending long amounts of time in a virtual world reduces the time for physical exercise and perhaps for having a healthy diet and enough sleep, all of which are also critical for the health of humans. Therefore, a balance between online and offline life is important, given that before humans also did not spend all their time with ikigai taishō activities.

It is important to reiterate or to clarify what AI-driven virtual hyper-personalised ikigai is not about. It is neither about virtual entertainment, nor about virtual love, nor about virtual religion. The first two are whole different businesses, while the latter one has been described elsewhere [30].

Virtual ikigai taishō activities, which lead to ikigai kan, must also be separated from addiction or wireheading. While ikigai taishō activities may have in common with addictive behaviour the persistent urge to engage in them, these activities neither have negative consequences, nor they cause harm as it is the case for addictions.

Wireheading is reward hacking to stimulate pleasure centres and has been described as a potential x-risk [31]. However, this would be the case if ikigai kan feelings can be reached without conducting ikigai taishō activities while this chapter focuses explicitly on ikigai taishō activities. Also, the potential extension mentioned above to link virtual hyper-personalised ikigai taishō activities with positive qualia would not be wireheading as it requires to execute the activity.

It is acknowledged that this chapter is partly speculative, which is the case for all research about the future, including AI safety research in general. Yet, due to their significance, i-risks must be considered as early as possible.

It is possible or even likely that humans would initially reject the outlined virtual ikigai taishō activities and consider them as dystopian. However, it has to be stressed that, also in the past, humans were often sceptical towards new technologies and then adjusted their behaviour. This applies also to recently emerged technologies, which our ancestors probably would have also considered as dystopian [e.g., 23]. Moreover, humans may be deterred by the belief that activities in virtual worlds are meaningless. However, it has been argued in philosophy that "there's no good reason to think that life in virtual reality will lack meaning and value. Nor is there reason to think its values will be limited to entertainment" [32, p. 312]. Therefore, if there are values in a virtual world, it should also be possible to find virtual ikigai taishō activities linked to these values.[6]

To sum up, it has to be again highlighted that AI-driven virtual hyper-personalised ikigais are neither dystopian nor meaningless but an opportunity. i-risks are serious albeit neglected issues considering developments in AI and other emerging technologies. All efforts to tackle i-risks are important and cannot be dystopian. In fact, the opposite, i.e., to continue to disregard i-risks, would be dystopian.

Therefore, this chapter aims to conceptualise approaches that make life worth living in times of advanced or transformative AI. It is envisioned that humans will have an online dashboard or platform where they are presented with AI-driven hyper-personalised virtual ikigai taishō activities, in which they can indulge as they wish.

Notes

1 See, e.g., https://www.gwern.net/Scaling-hypothesis.
2 This comes on top of the number of humans who have never found their ikigai and those who have found it but are too occupied to pursue it because of other obligations.
3 This reference is about uploaded humans but can to some extent also serve as an illustration for possibilities in virtual worlds without being uploaded.
4 There would be even no difference at all if we live in a simulation [28].
5 See [28] for the English translation.
6 If there were sentient digital minds, which is an open question, this would add a whole additional dimension of value in virtual worlds as indicated above.

References

[1] Kemp, N. (2022). *Ikigai kan*. Intertype Publish and Print: Melbourne.

[2] Ziesche, S. & Yampolskiy, R. V. (2020). Introducing the Concept of Ikigai to the Ethics of AI and of Human Enhancements. In *2020 IEEE International Conference on Artificial Intelligence and Virtual Reality (AIVR)*, 138–145. IEEE.

[3] Kamiya, M. (1966). *Ikigai-ni-tsuite[On 755 ikigai]*. Tokyo, Japan: MisuzuShyobou.

[4] Althaus, D., & Gloor, L. (2016). *Reducing risks of astronomical suffering: a neglected priority*. Berlin, Germany: Foundational Research Institute.

[5] Bostrom, N. (2002). Existential risks: Analyzing human extinction scenarios and related hazards. *Journal of Evolution and Technology*, 9. https://www.jetpress.org/volume9/risks.html

[6] Okuzono, S. S., Shiba, K., Kim, E. S., Shirai, K., Kondo, N., Fujiwara, T., ... VanderWeele, T. J. (2022). Ikigai and subsequent health and wellbeing among Japanese older adults: Longitudinal outcome-wide analysis. *The Lancet Regional Health-Western Pacific*, 21, 100391.

[7] Park, S. M., & Kim, Y. G. (2022). A Metaverse: Taxonomy, components, applications, and open challenges. *IEEE Access*, 10, pp. 4209–4251.

[8] Romero, A. (2022). DALL·E 2, Explained: The Promise and Limitations of a Revolutionary AI. towardsdatascience.com

[9] Kato, T. A., Kanba, S., & Teo, A. R. (2019). Hikikomori: multidimensional understanding, assessment, and future international perspectives. *Psychiatry and Clinical Neurosciences*, 73(8), 427–440.

[10] Deloitte (n.d.). Connecting with meaning. https://www2.deloitte.com/content/dam/Deloitte/ca/Documents/deloitte-analytics/ca-en-omnia-ai-marketing-pov-fin-jun24-aoda.pdf

[11] Zhu, J., & Ontañón, S. (2020). Player-centered AI for automatic game personalization: Open problems. In *International Conference on the Foundations of Digital Games*, 1–8.

[12] Ziesche, S. & Bhagat, K. (2022). *UNESCO State of the Education Report for India 2022 Artificial Intelligence in Education*. UNESCO: New Delhi.

[13] Choi, E., Bahadori, M. T., Schuetz, A., Stewart, W. F., & Sun, J. (2016). Doctor ai: Predicting clinical events via recurrent neural networks. In F. Doshi-Velez, J. Fackler, D. Kale, B. Wallace & J. Wiens (Eds.), *Machine Learning for Healthcare Conference* (pp. 301–318). PMLR.

[14] Hermann, E. (2022). Leveraging artificial intelligence in marketing for social good—An ethical perspective. *Journal of Business Ethics*, 179(1), 43–61.

[15] Pataranutaporn, P., Danry, V., Leong, J., Punpongsanon, P., Novy, D., Maes, P., & Sra, M. (2021). AI-generated characters for supporting personalized learning and well-being. *Nature Machine Intelligence*, 3(12), 1013–1022.

[16] Dellaert, B. G., Shu, S. B., Arentze, T. A., Baker, T., Diehl, K., Donkers, B., ... Steffel, M. (2020). Consumer decisions with artificially intelligent voice assistants. *Marketing Letters*, 31(4), 335–347.

[17] Griffiths, M. D., Kuss, D. J., & Demetrovics, Z. (2014). Social networking addiction: An overview of preliminary findings. *Behavioral Addictions*, pp. 119–141.

[18] Faggella, D. (2022a). You Don't Want What You Think You Want – AI and Procedurally Generated Worlds. https://emerj.com/ai-power/you-dont-want-what-you-think-you-want/

[19] Ziesche, S., & Yampolskiy, R. V. (2018). Towards AI Welfare Science and Policies. *Special Issue "Artificial Superintelligence: Coordination & Strategy" of Big Data and Cognitive Computing*, 3(1), 2.

[20] Ziesche, S., & Yampolskiy, R. V. (2019). Do No Harm Policy for Minds in Other Substrates. *Journal of Evolution and Technology*, 29(2), 1–11.

[21] Loosemore, R.P.W. (2014). Qualia Surfing. In R. Blackford & D. Broderick (Eds.), *Intelligence Unbound: The Future of Uploaded and Machine Minds*. Chichester: John Wiley & Sons, pp. 231–239.

[22] Bloch, L. R., & Lemish, D. (1999). Disposable love: The rise and fall of a virtual pet. *New Media & Society*, 1(3), 283–303.

[23] Faggella, D. (2022b). Your "Dystopia" is Myopia. https://danfaggella.com/dystopia/

[24] Hamada, H. T., & Kanai, R. (2022). AI agents for facilitating social interactions and wellbeing. *arXiv preprint arXiv:2203.06244*.

[25] Bostrom, N. (2003). Are we living in a computer simulation? *The Philosophical Quarterly*, 53(211), 243–255.

[26] Yampolskiy, R.V. (2018). Job ad: Universe Designers. In *Stories from 2045, Artificial Intelligence and the Future of Work*. In: C. Chase (Ed.), The Economic Singularity Club.

[27] Imai, T. (2012). The reliability and validity of a new scale for measuring the concept of Ikigai (Ikigai-9). *[Nihon koshu eisei zasshi] Japanese Journal of Public Health*, 59(7), 433–439.

[28] Fido, D., Kotera, Y., & Asano, K. (2020). English translation and validation of the Ikigai-9 in a UK sample. *International Journal of Mental Health and Addiction*, 18(5), 1352–1359.

[29] Yampolskiy, R.V. (2022). Metaverse: A Solution to the Multi-Agent Value Alignment Problem. *Journal of Artificial Intelligence and Consciousness*, 9(3), 1–11.

[30] Ziesche, S. (2019). An AI May Establish A Religion. In *Death and Anti-Death*, ed. C. Tandy, Volume 17, pp. 309–334. Ann Arbor: Ria University Press.

[31] Turchin, A. & Denkenberger, D. (2018). Wireheading as a Possible Contributor to Civilizational Decline. philpapers.org

[32] Chalmers, D. J. (2022). *Reality+: Virtual Worlds and the Problems of Philosophy*. Penguin UK.

5

Towards the Mathematics of Intelligence

Soenke Ziesche and Roman V. Yampolskiy

5.1 Set Theory of Minds

Since no common definition exists, we describe a mind as instantiation of intelligence.[1] This means that in addition to humans and many animals, non-animal intelligent agents constitute minds. It has been shown that the space of possible minds is vast, and attempts to characterise it have been made [e.g., 1–4]. In addition to a variety of minds—for humans inconceivable—grouped minds as well as nested minds may be possible.

As we will show below, set theory seems to provide tools to describe such constellations. In particular, we are looking at set union and complement as well as subsets. We will also make use of Yampolskiy's insight that the space of minds is countable and that all minds can be generated sequentially by a deterministic algorithm [1].

5.1.1 Grouped Minds: Set Union

Some scenarios of grouped minds appear frequently in our lives, while some are futuristic. Extensive research has been conducted on the collective behaviour of groups of humans as well as groups of animals (e.g., flocks of birds, schools of fish or ant colonies), especially regarding potentially emergent properties of such groups. Woolley et al. name one such property for groups of humans "collective intelligence" or "c factor", which they distinguish from "general intelligence", also known as "g factor" [5]. Woolley et al. show that the

> "c factor" is not strongly correlated with the average or maximum individual intelligence of group members but is correlated with the average social sensitivity of group members, the equality in distribution of conversational turn-taking, and the proportion of females in the group.
>
> [5]

DOI: 10.1201/9781003565659-6

This is contested by Bates and Gupta whose experiments (in a very controlled environment) instead revealed that higher individual IQ enhances the performance of a group, i.e., the group-IQ [6].

From our anthropomorphic viewpoint, we are used to singly embodied minds, but in addition, we can also conceptualise multiply and flexibly embodied minds. Goertzel provides a more general approach to grouped minds [7]. He defines what he calls "mindplex" as follows: it "is composed of a collection of intelligent systems, each of which has its own 'theatre of consciousness' and autonomous control system, but which interact tightly, exchanging large quantities of information frequently" and it "has a powerful control system on the collective level, and an active 'theatre of consciousness' on the collective level as well".

Heylighen coined the term "global brain" for the set of minds that is, since the emergence of the Internet, constituted of human minds as well as computers [8].

Sotala and Valpola analyse in more detail how "mind coalescence" as they call it could be achieved in a biological as well as post-biological way and what consequences it may have [9].

Yampolskiy et al. provide another novel approach of grouped minds in the form of a "Wisdom of Artificial Crowds" algorithm, which relies on a group of simulated intelligent agents to find solutions, which are autonomous and, in many cases, superior to individual solutions of all participating agents [10].

An open challenge for these constellations is to determine the threshold when those minds are merged into one as opposed to a set union of interacting minds.

5.1.2 Nested Minds: Subset

Nested minds are mostly speculation but relevant, especially since Bostrom analysed the likelihood that we are living in a computer simulation [11]. If this was true, this would be a constellation of nested minds, assuming that the simulating system has a mind too.

Other scenarios have been proposed; e.g., several by Kelly, such as "Global mind—large supercritical mind made of subcritical brains"; "Hive mind—large supercritical mind made of smaller minds each of which is supercritical"; "Low count hive mind with few critical minds making it up"; and "Borg—supercritical mind of smaller minds supercritical but not self-aware" [12].

Yampolskiy describes further scenarios of nested minds related to the artificial intelligence confinement problem [13]. Below we will explore also the third operation, complement of a set, i.e., subtracted minds, for which we did not find prior research.

5.2 Definition and Evaluation of Intelligence

Despite many attempts, none of the involved disciplines, such as psychology or artificial intelligence, has come up with a satisfying, mutually agreed upon definition of intelligence. Legg and Hutter (2007) provide an overview of the many definitions that have been proposed over the years and deliver eventually the general definition: "Intelligence measures an agent's ability to achieve goals in a wide range of environments" [14].

Although the first step, to agree on a unique definition of intelligence, has still not been completed, already extensive work has been conducted on the following step, which is to evaluate intelligence as a feature of psychometrics. Ability to evaluate intelligence is a precondition in order to be able to compare various intelligences, which is of interest regarding human as well as artificial intelligences. Hernández-Orallo worked extensively in this field and identified "three kinds of evaluation: human discrimination, problem benchmarks and peer confrontation" [15].

In this chapter, we use a definition of intelligence by Yampolskiy, which he developed within the framework of his "Efficiency Theory". There, he describes gaining of knowledge as finding and applying an efficient algorithm to an inefficiently stored information string. In other words, "Intelligence could also be defined as the process of obtaining knowledge by efficient means" [16].

Yampolskiy also states "historically, the complexity of computational processes has been measured either in terms of required steps (time) or in terms of required memory (space)" [16]. We suggest intelligence to be treated as a further dimension of fundamental computation resources in addition to time or memory. And while, for measuring and manipulating time and memory, there are established standards, we would also like to be able to measure or evaluate intelligence by its ability to turn certain information strings efficiently into knowledge.

Based on this, the scope of mathematics of intelligence of grouped and nested minds is to analyse and to compare the capabilities of all involved minds to turn information strings efficiently into knowledge and to establish relations between them.

5.3 Mathematics of Intelligence

We define minds by four parameters, the former two are mandatory, and the latter two optional: mind (x, i, a, t)

x (their unique identifier [1])
i (their intelligence, according to the "Efficiency Theory" [16])

a (their age, since i may vary depending on a)

t (their time of embodiment)

Based on the algorithm to generate sequentially all minds, Yampolskiy states, "A particular mind may not be embodied at a given time, but the idea of it is always present" [16]. Embodiment is the time period when a mind is exhibiting intelligence. This could the lifetime of living biological beings as we know them, but it could also be the period of implementation of a non-biological, artificial agent. This parameter is relevant since the intelligence of a mind may vary depending on the spacetime context of the embodiment, which is illustrated further below.

Now we look at various set operations and constellations, such as union, subset and complement.

5.3.1 Grouped Minds: Set Union

Of special interest is the question whether there is any scenario when the mind that results from combining other minds has a higher intelligence than all original minds? Such a phenomenon would be called emergentism. Another option would be that the intelligence of a grouped mind equals the highest intelligence of all minds in the union, which could be expressed by the MAX function.

Both are valid options. Normal IQ tests are timed and so having a number of minds working in parallel would increase the overall score, but at the highest level (160+), IQ tests are not timed and only a more capable mind would do better.

Intelligence is a computational resource like time, space or randomness. We propose arithmetic of intelligence to make it possible to perform analysis on this important resource. In particular, we concentrate on the minds at a sufficiently high level of performance; there, time is not an important factor, but overall ability to solve problems is the main feature in determining intelligence. Overall, arithmetic of intelligence functions is in a certain way similar to the Big O notation; there, one may disregard lower value terms. For example, for two intelligent systems with IQs equal to A and B, sum(A,B) = Max(A,B).

5.3.2 Nested Minds: Subset

Similarly of interest for the math of intelligences is the question whether in a nested constellation the mind in a subset could have a higher intelligence than the mind in the superset?

While intuitively it may seem that the mind in the subset is always less intelligent than the mind in the superset, this may not be the case given

Yampolskiy's definition of intelligence that it is all about better and more efficient algorithms to turn information into knowledge [16]. Assuming that, according to Plato's theory of Ideas [e.g., 17], algorithms exist independently and have just to be discovered rather than created, a scenario is imaginable that a mind in a subset happens to discover an efficient algorithm, which has not been discovered by the mind in the superset.

As mentioned above, one of the best known proposed nested constella- tions of minds are simulations. For such a constellation, this would mean that a simulated mind could find a better algorithm to turn certain informa- tion into knowledge than the simulating mind ever discovered, which would make the simulated mind more intelligent. Yet, the simulating mind would instantly get to know about the discovery of this algorithm (since the simu- lating mind by definition is copied on all processes in the simulation), i.e., would harness the intelligence of the simulated mind.

This means that the intelligence of a simulating mind can increase over time through two ways: 1) if the simulating mind itself finds better algo- rithms to turn information into knowledge or 2) if one of the simulated minds finds such better algorithms. At the same time, the intelligence of a simulating mind never decreases if we assume that a mind, which is intel- ligent enough to run simulations, must also be intelligent enough to keep a perfect memory.

Probably an inherent feature of (many)[2] simulations is that the subset of simulated minds changes over time through births and deaths. Based on the above insight, the intelligence of the simulating mind will be at least as high as the mind with the highest intelligence that ever inhabited the simulation, even if this mind has already passed away, since the simulating mind has access to all algorithms ever executed in the simulation.[3]

Therefore, it could be an incentive for a mind to simulate other minds as a tool to increase its own intelligence; hence, this may increase the likelihood that the second of the three propositions, which Bostrom uses in order to analyse the probability that our civilisation is currently living in a simula- tion, is *not* true: "Any posthuman civilization is extremely unlikely to run a significant number of simulations of their evolutionary history (or variations thereof)" [11]. Usually two motivations are named, as, e.g., by Tierney, why a posthuman civilisation would run simulations: research purposes or enter- tainment [18]. While research purposes are also about gaining knowledge, yet more through observation and monitoring, we are proposing here for the first time an additional motivation for running a simulation, which is to increase the intelligence of the simulating mind.

If this were the case, then a follow-up question would be whether a simu- lating mind could influence the simulation so that odds are higher that more intelligence is exhibited? Would this be through quality or quantity of the simulated minds? If through *quality*, then perhaps the simulating mind could focus on the simulation of other minds of whom it knows that their intelligence is above a certain threshold, e.g., 150:

$$\forall x : \text{mind}(x,i) \wedge i > 150$$

This would raise the question how the simulating mind could determine the intelligence of a simulated mind beforehand? We postulated above that the simulating mind might hope to benefit from increases of the intelligence of the simulated minds through discoveries of efficient algorithms. However, the simulating mind might have strategies to simulate minds that are promising to develop in this way.

If through *quantity* a simulating mind may try to maximise the number of simulated minds. One strategy to do so would be by somehow getting simulated minds to the level that they can run simulations themselves, i.e., by generating nested minds within nested minds.

The simulating mind could keep track of the simulated minds through their unique identifier and would aim, depending on its resources, to simulate as many as possible for a reasonable time frame and analyse how the intelligences of the simulated minds develop. Since also the time of the embodiment and the surrounding peer minds matter, the simulating mind could also experiment by successively simulating/embodying the same mind in different contexts and compare the performance of its intelligence.

Another thought is whether this would perhaps decrease the chances that we ourselves are in a simulation since so far our civilisation would have probably not contributed much to a potential simulating mind in order to increase its intelligence. It would seem to be a waste of time and resources to let run first a long evolution, instead of starting the simulation already with a fairly advanced civilisation [19]. Also, for efficiency purposes, such a simulating mind would probably run the simulated minds with a rather high pace to get results quicker. Yet our minds do not run particularly fast.

In this regard, we also have to consider nested minds within nested minds. For example, an advanced mind may have initiated a simulation with very intelligent minds in order to potentially increase its own intelligence as described above. Some of these simulated minds may have developed the capacity to run their own simulations, yet with less smart minds as they may have different capacities, intellectual as well as in terms of computational resources, or motivations to run simulations as the overall simulating mind. So, could we be one simulation on such a deeper level?

A special case would be if a simulated mind were encrypted. Then, the simulating mind would not be able to get access to its information. We have to distinguish between two sub-cases: 1) the simulated mind has the ability to encrypt some or all of its activities or 2) the simulating mind has encrypted the activities of the simulated mind. We shall not look here at the perhaps unlikely motivation why simulating minds may decide to encrypt their simulations.

Encrypted minds in a simulation would be the only scenario where a mind in the subset could have a higher intelligence than the mind in the superset. This would be the case when the mind in the subset discovers better algorithms to turn information into knowledge than the mind in the superset ever knew of, and the mind in the superset will not find out about it since it cannot decipher the encryption.

5.3.3 Subtracted Minds: Complement of a Set

Not much research has been conducted regarding the set operation "complement". Yet as part of the mathematics of intelligences, it is relevant to formalise how the intelligence of a grouped mind changes if certain minds are deducted from it.

There are two types of complements: absolute and relative. If A is a set, then the *absolute* complement of A is the set of elements, which are not in A. If A and B are sets, then the *relative* complement of A in B is the set of elements which are in B but not in A.

We look first at the *relative* complement: The deduction of one member or more (which equals A in the above definition) from a group of minds (which equals B in the above definition) can be visualised either as the ending of the embodiment of a particular mind, i.e., its death, or as a withdrawal from the group so that the intelligence of that particular mind does not contribute anymore to the overall intelligence of the group.

Let us look at the specific example whether and how the intelligence of humanity has changed after the deduction of Einstein's intelligence, which occurred on 18 April 1955, the day of his death:

$$\left\{ x_{\text{after}} \mid \text{mind}\left(x_{\text{after}}, i_{\text{after}} \right) \wedge \text{human\_mind}\left(x_{\text{after}} \right) = \text{true} \right\}$$
$$= \left\{ x_{\text{before}} \mid \text{mind}\left(x_{\text{before}}, i_{\text{before}} \right) \wedge \text{human\_mind}\left(x_{\text{before}} \right) \right.$$
$$= \text{true} \right\} \setminus \left\{ \text{mind}\left(m_{\text{Einstein}}, i_{\text{Einstein}} \right) \right\}$$

The question we are looking at is the relationship between i_{before}, the intelligence of humanity *before* Einstein's death, and i_{after}, the intelligence of humanity *after* Einstein's death.[4] As mentioned before, the time of embodiment and the level of development of the grouped mind are critical again. Since Einstein's work was very well documented and probably none of his relevant ideas got lost (unless he deliberately did not publish them), we may assume that after the deduction of his mind with the intelligence i_{Einstein}, the intelligence of humanity remained the same, i.e., $i_{\text{before}} = i_{\text{after}}$. In other words, the grouped mind has absorbed the intelligence of its former member Einstein. This was only possible, since the intelligence of the grouped mind was already on a certain level, while, e.g., an embodiment of Einstein in Stone Age would have taken place differently at least in the following two aspects: 1) the intelligence of Einstein would have almost certainly not reached the level it had reached in the 20th century, i.e.:

$$i_{\text{Stone Age}} \mid \text{mind(Einstein, } i_{\text{Stone Age}}, \text{Stone Age)}$$
$$< i_{\text{20th century}} \mid \text{mind}\left(\text{Einstein, } i_{\text{20th century}}, \text{20th century}\right)$$

The overall intelligence of humanity in Stone Age (or rather the community within which Einstein lived in this thought experiment since there was no global community at this time) would have been reduced after his demise, since the grouped mind was not at a level then to absorb his intelligence.

In general, for the *relative* complement, two points are relevant: 1) Did the deducted mind(s) make a significant contribution to the intelligence of the grouped mind? 2) If yes, was the intelligence of the grouped mind sufficiently advanced to harness the intelligence of these minds prior to their deduction? The latter point is also time-critical: e.g., if a human expert on a certain topic joins a group of humans for one hour only, this may not be enough time for the rest of the group to "pick that person's brain". Yet, for minds based on substrates that allow electronic exchange of information, one hour is likely to be sufficient to copy critical information from another mind that is connected to them for this time period only.

Moving on to the *absolute* complement, we would look at deducting minds from the universe of all possible minds. Basically, only two cases have to be distinguished, whether among the deducted minds is the one with the highest intelligence in the universe of all possible minds or whether that mind is *not* among the deducted ones.

In the first case, the overall intelligence of the complement will remain the same, which is the highest intelligence in the universe according to the Max function introduced above. In the second case, the Max function is also relevant because after the deduction of the mind with the highest intelligence, the intelligence of the complement equals the intelligence of the mind, which has the highest intelligence of all *remaining* minds.

5.4 Conclusion

We presented an introduction to the mathematics of intelligence. Our approach uses the set theory and then describes novel insights in set unions and complements as well as subsets of minds and the resulting intelligence. These findings can be seen as another contribution to the new field of "intellectology" [1]. Since this field of study is still in its infancy, also this work on mathematics of intelligence has to be considered as groundwork, which offers various routes for expansion.

In further research, we may look at other more complex set operations such as the symmetric difference, the Cartesian product or the power set and determine the resulting intelligence.

Notes

1 See below for a definition of intelligence.
2 There are also simulations imaginable without the anthropomorphic concepts of births and deaths, i.e., inhabited by a stable number of immortal minds.
3 If we look at the global brain approach by Heylighen, which was introduced above, there is the similar phenomenon that we preserve discoveries from ancestors, which means in an abstract way to keep particular algorithms over time [8]. Yet, at times, before printing and other information technologies were invented, there have been various instances of forgotten discoveries.
4 Also, many other humans passed away on 18 April 1955, but we assume their contribution to the overall intelligence of humanity to be negligible compared with Einstein's contribution.

References

[1] Yampolskiy, R.V. (2015) *Artificial Superintelligence: a Futuristic Approach*, Florida: Chapman and Hall/CRC Press.
[2] Sloman, A. (1984) "The Structure and Space of Possible Minds" In *The Mind and the Machine: Philosophical Aspects of Artificial Intelligence* (S. Torrance, ed.), Ellis Horwood LTD, pp. 35–42.
[3] Goertzel, B. (2006) *The Hidden Pattern: A Patternist Philosophy of Mind*, Boca Raton: Brown Walker Press.
[4] Yudkowsky, E. (2008) "Artificial Intelligence as a Positive and Negative Factor" In *Global Risk, in Global Catastrophic Risks* (N. Bostrom and M.M. Cirkovic, eds.), Oxford: Oxford University Press, pp. 308–345.
[5] Woolley, A. W., Chabris, C. F., Pentland, A., Hashmi, N., and Malone, T. W. (2010) "Evidence for a Collective Intelligence Factor in the Performance of Human Groups", *Science*, 330 (6004): 686–688.
[6] Bates, T.C., and Gupta, S. (2016) "Smart groups of smart people: Evidence for IQ as the origin of collective intelligence in the performance of human groups", *Intelligence*, 60: 46–56.
[7] Goertzel, B. (2003) "Mindplexes - The Potential Emergence of Multiple Levels of Focused Consciousness in Communities of AI's and Humans" in *Dynamical Psychology*. https://www.goertzel.org/dynapsyc/2003/mindplex.htm
[8] Heylighen, F. (2011) "Conceptions of a Global Brain: an historical review", *Evolution (Uchitel Publishing House)*, (1): 274–289.

[9] Sotala, K., and Valpola, H. (2012) "Coalescing Minds: Brain Uploading-Related Group Mind Scenarios", *International Journal of Machine Consciousness*, 4 (1): 293–312.

[10] Yampolskiy, R. V., Ashby, L., and Hassan, L. (2012) "Wisdom of Artificial Crowds – a Metaheuristic Algorithm for Optimization", *Journal of Intelligent Learning Systems and Applications*, 4 (2): 98–107.

[11] Bostrom, N. (2003) "Are You Living In a Computer Simulation?", *Philosophical Quarterly*, 53 (211): 243–255.

[12] Kelly, K. (2007) "A Taxonomy of Minds", Accessed October 1, 2018, https://kk.org/thetechnium/a-taxonomy-of-m/.

[13] Yampolskiy, R.V. (2012) "Leakproofing Singularity - Artificial Intelligence Confinement Problem" *Journal of Consciousness Studies*, 19 (1–2): 194–214.

[14] Legg, S., and Hutter, M. (2007) "Universal Intelligence: A Definition of Machine Intelligence", *Minds and Machines*, 17 (4): 391–444.

[15] Hernández-Orallo, J. (2016) *The Measure of All Minds*, Cambridge: Cambridge University Press.

[16] Yampolskiy, R.V. (2013) "Efficiency Theory: A Unifying Theory for Information, Computation and Intelligence", *Journal of Discrete Mathematical Sciences & Cryptography*, 6 (4–5): 259–277.

[17] Ross, W. D. (1951) *Plato's Theory of Ideas*, Oxford: Clarendon Press.

[18] Tierney, J. (2007) "Our Lives, Controlled From Some Guy's Couch" in *New York Times*, August 14, 2007, https://www.nytimes.com/2007/08/14/science/14tier.html.

[19] Yampolskiy, R.V. (2016) "On the origin of synthetic life: attribution of output to a particular algorithm", *Physica Scripta*, 92 (1): 1–10.

6

Designometry: Formalisation of Artefacts and Methods

Soenke Ziesche and Roman V. Yampolskiy

6.1 Introduction

Yampolskiy introduced the field of designometry, which aims to detect signatures of originators within artefacts. Owing to the diversity of artefacts, this type of research is currently pursued independently in different domains [1]. Therefore, Yampolskiy proposes an overarching approach through synergies and particularly through consolidation of methods of analysis from specific domains [1]. As he highlights, there could be a particular demand in the future to determine the originators of a specific type of artefacts, which are artificial or engineered life forms or minds.

In this chapter, we tackle this problem by presenting a survey of artefacts and a survey of designometry. In the survey of artefacts, we summarise existing definitions and ontologies, followed by an innovative approach to describe the space of artefacts by allocating identification numbers to them. In the designometry survey, we describe fields, which pursue designometry albeit not calling it that. Thereafter, we analyse the tools and methods of these fields and infer one abstract axiom as well as general heuristics for anyone trying to profile creators of artefacts, with a special focus on artificial minds. We also establish a link to the field of intellectology, which has been introduced by Yampolskiy [2].

6.2 Artefacts

6.2.1 Definition

Artefacts have been a topic in philosophy for a long time. Aristotle distinguished between things "that exist by nature" and those existing "from other

DOI: 10.1201/9781003565659-7

causes" [3]. For the latter group, he names a bed or a coat as examples and calls them "artificial products", since it requires an art of making things.

In the meantime, various other definitions for artefacts have been provided, which do not differ very much. For our purposes, the one by Hilpinen is sufficient: "An object is an artefact if and only if it has an author" [4, p. 156–157]. In the literature instead of author also creator, originator or agent is used.

6.2.2 Ontology

While artefacts have been described already for a long time as the definitions above show, formal ontologies have been developed only much more recently [5]. Around the end of the 20th century, research towards formal knowledge representation systems intensified, since this was required for various applications in the IT field. Borgo and Vieu present a detailed approach to extend formal ontologies for knowledge representation to include artefacts [6]. Among various existing ontologies, they propose that the Descriptive Ontology for Linguistic and Cognitive Engineering (DOLCE) is best suited for artefacts [7].

Borgo and Vieu added to existing categories in DOLCE the category of Physical Artefact, which became a subcategory of Physical Endurant, and thus a sibling of the categories Amount of Matter, Physical Object and Feature [6]. Furthermore, they made use of the quality feature in DOLCE by assigning all physical artefacts a single individual quality named capacity that characterises the capacities of the artefacts. This new quality enabled Borgo and Vieu to formalise a series of notions based on philosophical distinctions as well as common sense intuitions.

There are various ways to categorise artefacts according to their qualities. As this is a wide topic of philosophical research [8], here only the most important categories are mentioned: an example for a singular, concrete object would be the Eiffel Tower in Paris. If the simulation of universes turns out to be feasible, then a simulated universe would be an extreme example for a singular, concrete object. As opposed to singular objects, which are unique, there are also artefacts, which have more than one or many instances. The instances of such artefacts have the same blueprint and the same characteristics. An example would be paper clips. A particular instance of this type of artefact would be in this example a specific paper clip. As opposed to concrete objects, artefacts can also be abstract. An example would be a text or any object in a digital environment. The above categories can be divided further in subcategories. For example, singular, concrete objects may be independent or dependent. An example for an independent object would be a chair. As opposed to this singular, concrete dependent objects cannot exist without a substrate. An example would be a tunnel under a mountain. This object cannot be detached from that mountain. In addition to numerous other criteria such as the method of manufacture, material properties or the intended use of the artefact, it can also be distinguished if an artefact has one author or more than one author (collectively produced artefacts) or whether the artefact is intended by the author or unintended. For example, when a tailor makes a coat for his customer, his intention is to make a coat of a certain size and style, but he also produces scraps of cloth as by-products of her work. Moreover, there are also biological artefacts, which are based on DNA sequences that are not found in nature, i.e., artificial recombinant molecules.

6.2.3 Space of Artefacts

As shown above, there have been various approaches to classify artefacts, yet there appears to be no attempt to describe the space of artefacts by allocating artefacts identification numbers.[1]

Every ever-produced artefact can be assigned a unique identification number based on two parameters: the location[2] coordinates and time where and when the creator initiated the production of the artefact. (This also includes then unfinished or "work-in-progress" artefacts as they are already considered artefacts once the production has begun.) Already these two parameters allow for unique identification of artefacts, including those, which have many instances, e.g., paper clips. Even in factories where a large number of paper clips are produced, the location and time of the beginning of their production differs if the coordinates are sufficiently fine-tuned.

For dependent artefacts, the location never changes, e.g., the Eiffel Tower, while others are moved around, yet for identification, the location coordinates are used where its creation began. In the same way, unique identification numbers can be assigned to abstract artefacts. Obviously, there are many location–time pairs, which do not represent artefacts (i.e., every pair, where and when no creation of an artefact began).

Yampolskiy suggested that the Kolmogorov Complexity measure could be applied to identification numbers representing mind designs, e.g., to determine whether shorter representations of the original number are possible [2]. The same idea could be transferred to the identification numbers of artefacts.

Based on this approach of location–time pairs, it can be inferred that the set of identification numbers of artefacts is finite.

As shown, these two parameters (location and time) allow already for unique identification of artefacts, yet a third interesting parameter would be the creator of the artefact. Creators of artefacts have always a mind (which distinguishes artefacts from other objects created by nature), and Yampolskiy has shown that the space of potential minds is infinite but countable, which enables us to use the identification number of a mind as third parameter and define the space of artefacts as follows [2]:

$$\text{Space of artefacts} = \{m, t, l \mid \text{A mind } m \text{ initiated the production of an artefact at the time } t \text{ and the location } l\}$$

As we now have three parameters, also the possible subsets with the cardinal number 2, i.e., creator–location and creator–time, allow for unique identification of artefacts. Similarly, there are many creator–location and creator–time pairs, which do not represent artefacts.

For various reasons and within different sciences, it is of interest to uncover the time of creation, the location of creation and/or the creator of an artefact. The particular process to identify the creator of an artefact has been called "designometry" by Yampolskiy [1].

6.3 Designometry Survey

6.3.1 Definition

Yampolskiy introduced designometry as a subfield of intellectology [1, 2, 9] and describes it as the field of study, which aims "to uncover a 'signature' of the originator in the artefact and from it to identify the agent responsible or to at least learn some properties, of the design process, which produced the artifact". Yampolskiy highlights that designometry could become particularly relevant for potentially engineered life or minds [1]. Looking at the creator–location–time triple defined above to identify artefacts, designometry is exploring ways to retrieve the first parameter of the triple, i.e., the creator.

The finite set of all artefacts can be split into the two subsets of all those artefacts whose creators can be identified and those artefacts whose creators *cannot* be identified. However, this subdivision depends on two more parameters, which are the types of minds [2] who are pursuing the designometry and the times of their embodiment. Certain types of minds may have the capacity/intelligence to successfully identify the originators of certain artefacts after a certain time of embodiment, involving education, research, etc., while this may be beyond the capacity/intelligence of other minds.

Essentially, designometry is a two-stage process, which looks successively at the following two questions:

1) Is an object an artefact?

2) If yes, is it possible to identify the mind who created this artefact?

Within various sciences, the quest for the creator of artefacts is part of the research—a fact that motivated us to introduce classifications of artefacts above. The type of artefact differs depending on the particular science, and methodologies are usually not coordinated among the sciences. It is an aim of designometry to create synergies so that sciences could benefit from each other in their studies. Therefore, in what follows we survey all these fields in which relevant work is done and try to abstract away details about the field, while keeping the methods in an effort to extract generic methods of designometry. As we will see, some of these fields focus only on the second question above, since by definition of their domain, they deal with artefacts only, e.g., stylometry, while for other domains, both questions are relevant, e.g., archaeology.

6.3.2 Archaeology/Archaeometry

Archaeology is the classic science dealing with artefacts. Archaeology is restricted to artefacts produced by humans and deals mostly with concrete artefacts, since over 99% of the human development has occurred before abstract artefacts such as written texts or digital objects existed. Different phases can be distinguished: a significant amount of time and resources of an archaeological investigation are dedicated to survey areas of interest and to uncover artefacts, often through excavation. Only in the subsequent phase, the artefacts are analysed, for which methods of archaeometry are applied. Sophisticated techniques have been developed to determine the time of creation as well as the function of unknown artefacts, while the identification of a specific creator of an ancient artefact is in many cases neither possible nor pursued. An example for uncovering a signature within archaeometric research is provided by Labati et al. for the specific case of latent fingerprints on clay artefacts [10]. To distinguish this domain from the other ones below, it can be stated that archaeology and archaeometry do neither deal with biological artefacts nor concentrate on abstract ones, which are both of interest for designometry.

6.3.3 Artimetrics

Yampolskiy and Gavrilova introduced the innovative field of artimetrics as an extension to the known domain of biometrics [11]. In addition to be able to identify, classify and authenticate biological entities through sensors, which is the field of biometrics, the need arose to be able to do the same with their virtual representatives, i.e., embodied as well as virtual robots, software, and virtual reality agents. The field of artimetrics aims to identify such artificial entities based on their outputs or behaviour. In the ontology above, it was mentioned that artefacts could be abstract or virtual.

6.3.4 Astrobiology

Astrobiology is a science, which, among other things, examines whether extant or extinct life in the universe exists or existed, of which we are not aware, and if yes, through which methods it can be detected. One established yet up to now unsuccessful method is to monitor electromagnetic radiation for signs of transmissions from other life outside earth. To complement these efforts, Freitas proposed the Search for Extraterrestrial Artefacts based on his "Artifact Hypothesis: A technologically advanced extraterrestrial civilisation has undertaken a long-term programme of interstellar exploration via transmission of material artifacts" [12]. He distinguishes four classes of potential artefacts: astroengineering, self-replicating artefacts, passive artefacts and active probes. In this regard, astrobiology would be an example where both issues above are relevant: first, objects have to be identified as artefact, which is potentially followed by the quest for the creator. However, Haqq-Misra and Kopparapu show that despite long-lasting searches, non-terrestrial artefacts would likely remain not be detected because of the vastness of the universe [13]. Wright suggests considering "a prior indigenous technological species" in our solar system; thus, earth or nearby planets would have to be scrutinised for artefacts of such species [14].

6.3.5 Behavioural Biometrics

Due to the proliferation of interaction between humans and electric devices, the need for various authentication methods increases. In addition to often used unique physiological characteristics of a user, another category is unique behavioural characteristics. The latter have been surveyed by Yampolskiy and Govindaraju and can be divided into behaviours that produce artefacts, e.g., texts, emails, sketches or programming codes, and those which do not produce outputs, e.g., car driving style, game strategies, lip movements or mouse dynamics [15]. These artefacts in the first category may include unique signatures of the user, which can be further analysed through stylometric methods, which are introduced below.

6.3.6 Forensics Science

Forensic science concerns the collection and analysis of evidence linked to a criminal investigation. A subset of the evidence could be artefacts, e.g., an exploded or unexploded device, in which case the identification of the creator may advance the criminal investigation.

The subfield digital forensics addresses evidence found in digital devices, which becomes more and more relevant. All such evidences are abstract artefacts. Methods to identify the authors of such digital content are described below under "Code stylometry".

6.3.7 Stylometry

Stylometry is a domain, which initially focused on the particular artefact of written texts, i.e., the identification of an author of an anonymous or disputed text. There it is sometimes called authorship attribution. This field has progressed recently a lot due to the availability of a large digital text corpus, which can be utilised for statistical analysis. Moreover, stylometry has been extended to other areas of creative artefacts. Backer and van Kranenburg deliver an approach for the subfield of musical stylometry based on statistical pattern recognition [16]. Various approaches also tackle the subfield of visual stylometry [e.g., 17–19]. The prime motive is often to find methods to authenticate songs and paintings, respectively, and at the same time to uncover forgery.

6.3.7.1 Code Stylometry

Code stylometry is the subfield, which aims to find methods to de-anonymise the creator of programming codes, which is also motivated to improve detection of plagiarism. Code stylometry is particularly interesting, since automated methods through machine learning have been developed which can be applied to a large code corpus. Caliskan-Islam et al. examine machine-learning methods to identify source code authors of C/C++ using coding style [20]. A distinction has to be made whether source code or merely compiled binary code is available for analysis. Binary code is much harder to de-anonymise, which was tackled, e.g., by Rosenblum et al. [21] and further expanded by Caliskan-Islam et al. [22].

6.3.8 Simulation Detection

Bostrom analysed the likelihood that we are living in a computer simulation [23]. If this were true, our universe itself would be an artefact (and the set of naturally occurring things in our universe would be empty), and the identification of its creator/simulator would be of high scientific interest, yet extremely difficult if not impossible. Due to these challenges, there are not many (scientific) attempts in this regard. For example, Hsu and Zee argue that cosmic microwave background could be used as a communication channel with a potential creator/simulator [24]. Beane et al. take a different approach and aim to show that a simulator could in principle detected because of the finiteness of resources [25]. Both these methods of designometry would differ from all the other methods listed here.

6.3.9 Synthetic Biology

Synthetic biology is an interdisciplinary field, which is still in early stages and examines the design and construction of new biological entities or the redesign of existing biological entities [e.g., 26]. For various reasons, it can be of interest based on the resulting biological artefact to identify its creator. The creator may have placed, e.g., intentionally a signature in the DNA, which is called steganography [27] and is the practice of concealing information within unsuspicious cover carriers. Beck et al. provide an approach to find such messages and thus possibly identify the author [28]. If the creator has left no such message, potentially methods from code stylometry could be applied, i.e., looking for specific patterns in the synthetic code to reveal the author's identity.

6.4 Designometry Methodologies

In this section, we aim to present generic methods for designometry, either by deriving them from the individual methods applied by the above fields, which use designometry, or by introducing original techniques. These methods constitute of one axiom and a set of heuristics. As described, designometry is a two-stage process, each stage using different approaches, which is reflected in the structure of this section.

6.4.1 Is an Object an Artefact?

Some of the fields above have to address this question first, e.g., archaeology, while others do not have to since by definition they exclusively deal with artefacts, e.g., stylometry. As described above, even for biological samples, this question may be relevant and they may turn out as artefacts. While synthetic biology deals with the manufacture of new biological entities, the reverse question "how to separate engineered from natural biological entities" has not received a lot of attention in the literature of the field. Yampolskiy stresses that precisely this issue could become very important in the future because of expected advances in synthetic biology, which may, however, lead to unforeseeable and undesirable consequences [1].

We propose the following axiom to determine whether an object is an artefact.

Axiom 1
An object is an artefact if it contains writing.

We define writing as a means of communication that uses signs. Therefore, if written signs can be found in an object, we conclude that the object is an artefact. Signs are studied in the field of semiotics. We use here the dyadic[3] approach by Saussure, which distinguishes between a signifier, i.e., the form a sign takes, and the signified, i.e., the concept a signifier refers to [30]. Written signs can be iconic or symbolic.[4] Iconic signs are characterised by a similarity between signifier and signified. An example is a drawing of a chair. Symbolic signs are arbitrary, and the relation between signifier and signified is conventional within a particular language. An example is the writing of the word "chair". This means symbolic signs can be anything as long as there is a social convention about its meaning.[5]

Symbolic signs, which are the main subject of Saussure's research, create for the above axiom the challenge to identify writing in an artefact. This challenge would be particularly hard for artefacts, which were created by a type of mind whose written language humans do not know. Using Saussure's terms, the twofold challenge is not only to get an understanding of the signifier, but before to determine whether any material within an artefact constitutes at all a signifier.

Ways to tackle these challenges are explored in the subfield exosemiotics. Exosemiotics studies signs that theoretically could be understood by other minds [e.g., 31]. An early attempt is the constructed language Lincos, which Freudenthal created to be understandable by any possible intelligent mind [32]. Yet the space of possible minds is vast [e.g., 2] and so could be potentially their respective use of signs.

Once an object has positively been declared to be an artefact, the second designometric query has to be tackled, which we approach by the following heuristics:

6.4.2 Is It Possible to Identify the Mind Who Created This Artefact?

6.4.2.1 Heuristic b.1

One option to identify the mind who created the artefact is to look for an *intentional* signature.

This heuristic is relevant for artefacts where the creators intentionally included writing, which allows backtracking to the creator. In trivial cases, these could be literally signatures or other physical or digital watermarks, which serve for copyright protection. Yet these signatures could also be forgeries, which is another field of research. Intentional signatures on artefacts are relevant in forensic science as they may serve as claim of responsibility for a misdoing. One example would be illegal graffiti with so-called tagging as a form of personal expression. Likewise, forensic scientists have to be aware of forgeries. A special case is intentional but hidden signatures, which was introduced above as steganography and is explored, e.g., in synthetic biology.

In some contexts, the identification of a species as creator instead of an individual is already sufficient. This is the case quite often in archaeometric research, but also in astrobiology, it would be an unprecedented success if an artefact with a signature of a species not originating from earth could be found. An example for such an artefact would be the so-called Pioneer plaques, which travelled to space on board the Pioneer 10 and Pioneer 11 space probes in 1972 and 1973, respectively. These plaques contain a pictorial message, based on exosemiotic considerations, for a potential encounter with extraterrestrial life, for whom it is intended to be sufficient to identify the human species as creator of this artefact rather than the human individuals involved in the production of the plaques.

6.4.2.2 Heuristic b.2

One option to identify the mind who created the artefact is to look for an *unintentional* signature.

While intentional signatures often signal that the creator wants to be linked to this artefact and no scientific work for identification is required, the situation is different when the creator obviously does not want to be identified and searching for unintentional signatures is used as a method to do so.

This heuristic is even more common in forensic science than b.1, since usually culprits do not want to be identified. Therefore, methods of forensic science focus on the revelation of unintentional signatures at artefacts linked to a crime such as fingerprints or more recently DNA. The search for signatures is not a priority within archaeometric research, but one example by Labati et al. about latent fingerprints on clay artefacts was mentioned above [10].

Other types of unintentional signatures would be certain patterns, which could be unambiguously retraced to the creator. Two categories can be distinguished: whether it is attempted to detect the pattern from the design and/ or the behaviour of the artefact or from the underlying code, which defines the artefact.

Examples for the first category would be artimetrics and behavioural biometrics. Here, a black box approach is applied and the search for patterns focuses on the observed properties, output or behaviour. Another example is the nests of weaverbirds. Bailey et al. describe how nests can be attributed to individual weaverbirds through computer-aided image texture classification [33]. The signatures have to be considered unintentional if we assume that the minds of the creators are not sufficiently sophisticated. Also, in certain contexts, the identification of a species through unintentional signatures instead of an individual is already scientifically satisfying. This is the case for many animal-built structures [34].

An example for the exploration of patterns in the code would be code stylometry. In the context of biological entities, the distinction above could be formulated as to search for a signature in the phenotype or in the genotype of the organism. Overall, the availability of a code or genotype is preferred, since this allows performing more precise analysis of a specific dataset.

If the desirable code is not at hand, another tool related to designometry becomes relevant, which is reverse engineering. This is the process of taking an artefact and trying to obtain knowledge about its construction plan or its code. Reverse engineering can be hard and costly. Villaverde and Banga provide an overview of available reverse engineering methods for biological entities, which is relevant, since our special interest lies in designometry related to engineered life [35].

Therefore, reverse engineering can be seen as a function from the space of artefacts, represented by the set of identification numbers of artefacts, to the code of the artefact, whereby the artefacts are represented by its identification number:

$$re : \{artefacts\} -> \{code\}$$

Artefacts with more than one instance have individual identification numbers as defined above but share the same code; thus, this function is not bijective.

6.4.2.3 Heuristic b.3

One option to limit the set of minds who created the artefact is to focus on those minds who have a motive, goal or drive to create this artefact (and to potentially sign it).

Artefacts are created for a reason; thus, the potential creators of a particular artefact can be narrowed down to the subset of minds that have a motive, goal or drive to create this artefact. In forensic science, potential motives for a crime play an important role. According to Bostrom's orthogonality thesis, the goals and the intelligence levels of minds are independent of each other [36]; hence, the heuristics b.3 and b.4 are independent. Regarding artefacts, which are created by artificial minds, Omohundro [37, 38] defines general drives for such activities: efficiency, self-preservation, acquisition and creativity. This heuristic addresses also potential writing on the artefact, which is relevant for the above axiom and heuristics, as this may reveal a sub-goal by the creator to not only produce the artefact but also to sign it.

6.4.2.4 Heuristic b.4

One option to limit the set of minds who created the artefact is to focus on those minds who have the necessary skillset, education and intelligence to create this artefact.

Apart from artefacts, which minds can create only because of their genetic predisposition, it requires intelligence as well as education to create certain artefacts, which only a subset of minds acquires. For example, for a long time, only Chinese and Japanese had the knowledge and skills to produce porcelain, while the production of European porcelain only started in 18th century. Therefore, the set of creators of any porcelain produced before 1700 can be limited to Chinese and Japanese minds.

The minds with a certain skillset to create certain artefacts could be subsets within a species, but also whole species. Above we introduced animal-built structures, and, e.g., if we look at the artefact spider webs, it has many instances, but the set of potential creators is limited to the set of spiders.

It has to be noted that the creator does not have to be the inventor of a particular artefact but has acquired the skills to produce the artefact. The inventor of an artefact with more than one instance could be identified by checking which creator–location–time triple of all instances of this artefact has the lowest/earliest time number and, thus, to determine when this artefact was produced for the first time.[6]

The degree of skills required to produce an artefact varies highly. It can be inferred that the more sophisticated the artefact is, the smaller is the subset of minds who is able to create this of artefact. As designometry aims to focus on sophisticated artefacts such as artificial minds in particular, this heuristic could be useful.

6.4.2.5 Heuristic b.5

One option to limit the set of minds that created the artefact is, if the time and the location coordinates when and where the production of the artefact was initiated can be determined, to look only at minds who were embodied at this time and were able to reach this location.

As established above, all artefacts can be defined through a creator–location–time triple, and designometry addresses the quest for the creator. However, the other two parameters could be supportive as this heuristic shows.

This heuristic is, e.g., applied in archaeology and also in forensic science. While in archaeometric research, it is often satisfying to limit the set of creators to a certain epoch, in forensic science, the manhunt is for specific individuals. Therefore, the whereabouts of suspects in relation to artefacts, which are linked to the investigation, constitutes critical information.

6.4.2.6 Heuristic b.6

One option to identify the mind who created the artefact is to look for witnesses, testimonies, coverage, footage, logfiles or sensor data of the creation of the artefact.

According to this heuristic information about the creator of an artefact could be gathered if another mind or another sufficiently equipped artefact has witnessed the creation of the artefact. On this heuristic, we have to rely especially if the artefact does not exist anymore. One example would be the Seven Wonders of the Ancient World, of which nowadays only the Great Pyramid of Giza still is in existence. That the other six also existed and that, e.g., the creator of the Statue of Zeus at Olympia was a Greek sculptor named Phidias we have to believe because of testimonies.

For testimony by humans, the criterion of credibility is relevant. However, more and more features and activities in the physical world are captured electronically, e.g., through sensors. These data naturally include also evidence about creation of artefacts. This could be relevant for the creation of artefacts, which involves computers such as artificial minds, as logfiles or similar digital traces may link to the creator.

Although the introduced axiom and heuristics mark an early phase of designometry, the envisaged contributions of designometric methods can be summarised as follows:

> To provide synergies to fields, which have the overlapping sub-goal to identify creators of artefacts but which did not cooperate much yet and enhance as well as consolidate their efforts.

> To provide time-critical input to AI safety from an innovative angle as the heuristics are applicable to the subset of artificial minds.

6.4.3 Designometry as a Sub-Branch of Intellectology

As Yampolskiy suggested, designometry can be seen as a subfield of intellectology [1], which was introduced by Yampolskiy in order to examine in more detail features of the space of minds [2]. More precisely, designometry can be seen as a function from the space of artefacts to the space of minds.

$$d : \{\text{artefacts}\} -> \{\text{minds}\}$$

Since we have shown that artefacts can have identification numbers assigned to and Yampolskiy has done the same for the space of minds, this is a function between natural numbers [2]:

$$d : \mathbb{N} -> \mathbb{N}$$

This means that the designometry function maps the unique identification number of an artefact to the unique identification number of the mind of its creator. A special case is the category "collectively produced artefacts", which was introduced above. These artefacts have more than one creator, i.e., the identification number of the artefact is mapped on a set of numbers, which are the identification numbers of all its creators.

As shown above, an intermediate stage is often (but not always) to map the artefact to its code, in which then hints for the mind, who created the artefact, may be found.

The introduction of the creator–location–time triple enables us to define further interesting sub-categories. Creator–time pairs for artefacts are only possible during the time period when the creator with this particular mind has been embodied—a fact that is utilised in heuristic b.5. An interesting set is, e.g., the set of all artefacts produced by a particular mind during its embodiment.

Of particular relevance is the set of artefacts that comprises all artificial minds, which can also be referred to as the set of not naturally occurring minds within the space of possible minds, and the question which minds are capable to create them (see heuristic b.4)? In other words, for the set of artificial minds, what is the set of potential creators within the creator–location–time triple?

Artificial minds constitute the intersecting set of the set of minds and the set of artefacts. Since artificial minds may also produce other artefacts, a nested constellation emerges. This raises further interesting questions: what is the set of artefacts that can only be produced by artificial minds, but not by naturally occurring minds? Are there artefacts, which can only be produced by those artificial minds, which themselves have been produced by artificial minds? This question can be applied to deeper levels of nesting too. In this regard, it makes probably sense to distinguish between offsprings of artificial minds resulting from reproduction, which are by definition also artefacts, and those artefacts, which are produced by artificial minds outside of reproduction. One trivial conclusion is that the time parameter of the creating mind must be lower than the time parameter of the created mind.

Through the introduced methodologies of designometry, various sets and subsets of artefacts can be defined as partly illustrated in Figure 6.1.[7]

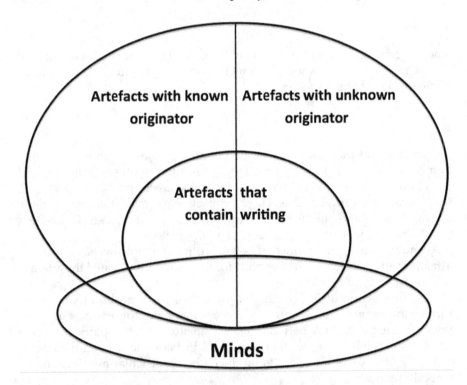

Figure 6.1 Categories of artefacts including artificial minds.

6.5 Conclusion and Future Work

We have presented two surveys, a survey of artefacts and a survey of designometry. We demonstrated how these surveys are interconnected. The new field of designometry aims to find general tools and methods to identify the creator of artefacts. Currently this field is divided into specialised subfields for particular purposes, e.g., criminal investigation, or particular artefacts, e.g., fine art.

We aimed to provide a bigger picture by specifying the space of artefacts through a creator–location–time triple and extracted the two-stage process that first it has to be ascertained that an object is an artefact and then the quest for the creator has to be tackled. For both stages, we have formulated an axiom and general heuristics, some of which were partly and rather implicitly applied before and some of which are innovative. We have determined that writing on an object is a clear indicator that the object is an artefact. For the identification of the creator the examination of the code for signs appears to be more promising than scrutinising properties or behaviour of an artefact. Therefore, procedures such as reverse engineering to obtain the code are a relevant instrument for designometry.

Yampolskiy has proposed the field of designometry with an outlook towards a particular subset of artefacts, which are artificial minds [1]. Given the ongoing progress in the concerned technologies, such research is very timely. Methods to find out the creator of artificial minds are likely to be relevant for several reasons, e.g., for proper registration, but can be seen in particular as a contribution to the field of AI safety, which entails the identification of originators of malicious systems as a critical step to curtail such systems if possible. Both the specification of the space of artefacts as well as the proposed axiom and the initial set of heuristics to identify creators of artefacts can be seen as groundwork for future AI safety research.

Regarding future work, in addition to the proposals and open questions, which were mentioned above, we may look at a fourth parameter to define the space of artefacts, supplemental to creator, time and location, which would be the goal or purpose of the particular artefact. In heuristic b.3 above we looked at the goals of the creators of artefacts, which could be connected to the goals of the artefacts, referred to by Tegmark as outsourcing of goals through engineering [39]. The feature "intended use of the artefact" has been introduced before within the ontology of artefacts, but it had not been linked to AI safety. In addition to establishing who created artefacts and when and where, it is many contexts important to know whether the artefact has benevolent or malevolent goals.

Tegmark highlights the increasing relevance of goals of artefacts, since in addition to living organisms having goals a "rapidly growing fraction of matter was rearranged by living organisms to help accomplish their goals" [39]. He also presents data that show that "most matter on Earth that exhibits goal-oriented properties may soon be designed rather than evolved". This interesting observation motivates us to expand our work towards goals of artefacts.

Finally, it has to be reiterated as by Yampolskiy [1] that when we discuss engineered life and artificial minds, we do not support by any means non-naturalistic notions be it god(s), creationist myths or religion, but merely the engineering of biological entities or minds in other substrates.

Notes

1 Similarly, Yampolskiy has described the space of possible mind designs [2].
2 The specification of the location is not topic of this chapter, but it should be as precise as possible. This means, for artefacts produced by nanotechnology, the location would be measured on an atomic and molecular scale.
3 An alternative in semiotics is the triadic approach by [29].
4 Another category is signs that are indexical. These are signs where the signified causes the signifier. For example, fire causes smoke; in other words, smoke signifies fire. However, this category does not apply to written signs.
5 In semiotics, it is usually argued that also the grasping of the meaning of iconic signs involves to some extent social conventions.
6 An exceptional case is artefacts that were invented more than once independently at different times and dates. An example is the calculus, which is credited to both Newton and Leibniz.
7 The sizes of the areas do not represent proportions of their cardinality.

References

[1] Yampolskiy, R. V. (2016). On the origin of synthetic life: attribution of output to a particular algorithm. *Physica Scripta*, 92(1), 013002. https://www.researchgate.net/profile/Roman_Yampolskiy/publication/310736484_On_the_origin_of_synthetic_life_Attribution_of_output_to_a_particular_algorithm/links/5836117708aec3fe331c5203/On-the-origin-of-synthetic-life-Attribution-of-output-to-a-particular-algorithm.pdf

[2] Yampolskiy, R. V. (2015). The space of possible mind designs. In *International Conference on Artificial General Intelligence* (pp. 218–227). Springer International Publishing. https://www.researchgate.net/profile/Roman_Yampolskiy/publication/300646188_The_Space_of_Possible_Mind_Designs/links/5737f1b408ae9ace840bfa3a.pdf

[3] Aristotle, Physica, in *The Works of Aristotle Translated into English* (Volume II), David Ross (ed.), Oxford: Clarendon Press, 1930.

[4] Hilpinen, R. (1993). Authors and artifacts. In *Proceedings of the Aristotelian Society* (Vol. 93, pp. 155–178). Aristotelian Society, Wiley.

[5] Franssen, M., Kroes, P., Reydon, T., & Vermaas, P. E. (Eds.). (2013). *Artefact kinds: Ontology and the human-made world* (Vol. 365). Springer Science & Business Media.

[6] Borgo, S., & Vieu, L. (2009). Artefacts in formal ontology. *Handbook of Philosophy of Technology and Engineering Sciences, 9*, 273–307.

[7] Masolo, C., Borgo, S., Gangemi, A., Guarino, N., & Oltramari, A. (2002). Wonderweb deliverable d17. *Science Direct Working Paper No S1574-034X (04)*, 70214-8. https://www.loa.istc.cnr.it/old/Papers/DOLCE2.1-FOL.pdf

[8] Hilpinen, R. (2011). Artifact. In E. N. Zalta (ed.), *The Stanford Encyclopedia of Philosophy*. https://plato.stanford.edu/entries/artifact/

[9] Ziesche, S., & Yampolskiy, R. V. (2017). High performance computing of possible minds. *International Journal of Grid and High Performance Computing (IJGHPC), 9*(1), 37–47.

[10] Labati, R. D., Genovese, A., Piuri, V., & Scotti, F. (2012). Two-view contactless fingerprint acquisition systems: A case study for clay artworks. In *Biometric Measurements and Systems for Security and Medical Applications (BIOMS), 2012 IEEE Workshop on* (pp. 1–8). IEEE. https://piurilabs.di.unimi.it/Papers/bioms_2012.pdf

[11] Yampolskiy, R. V., & Gavrilova, M. L. (2012). Artimetrics: Biometrics for artificial entities. *IEEE Robotics & Automation Magazine, 19*(4), 48–58.

[12] Freitas, R. A. (1983). The search for extraterrestrial artifacts(SETA). *British Interplanetary Society, Journal (Interstellar Studies), 36*, 501–506. https://www.rfreitas.com/Astro/SETAJBISNov1983.htm

[13] Haqq-Misra, J., & Kopparapu, R. K. (2012). On the likelihood of non-terrestrial artifacts in the Solar System. *Acta Astronautica, 72*, 15–20. https://arxiv.org/pdf/1111.1212.pdf

[14] Wright, J. T. (2017). Prior indigenous technological species. *International Journal of Astrobiology*, 1–5. https://arxiv.org/pdf/1704.07263.pdf

[15] Yampolskiy, R. V., & Govindaraju, V. (2008). Behavioural biometrics: A survey and classification. *International Journal of Biometrics, 1*(1), 81–113. https://pdfs.semanticscholar.org/1958/eebd997a2e90b88e1f8bb5345ec88408a1ce.pdf

[16] Backer, E., & van Kranenburg, P. (2005). On musical stylometry—a pattern recognition approach. *Pattern Recognition Letters, 26*(3), 299–309. https://www.sciencedirect.com/science/article/pii/S0167865504003393

[17] Hughes, J. M., Mao, D., Rockmore, D. N., Wang, Y., & Wu, Q. (2012). Empirical mode decomposition analysis for visual stylometry. *IEEE Transactions on Pattern Analysis and Machine Intelligence, 34*(11), 2147–2157. https://ieeexplore.ieee.org/abstract/document/6127875/

[18] Qi, H., Taeb, A., & Hughes, S. M. (2013). Visual stylometry using background selection and wavelet-HMT-based Fisher information distances for attribution and dating of impressionist paintings. *Signal Processing, 93*(3), 541–553. https://www.sciencedirect.com/science/article/pii/S0165168412003623

[19] Jacobsen, C. R., & Nielsen, M. (2013). Stylometry of paintings using hidden Markov modelling of contourlet transforms. *Signal Processing, 93*(3), 579–591. https://people.math.aau.dk/~mnielsen/reprints/2013_stylometry.pdf

[20] Caliskan-Islam, A., Harang, R., Liu, A., Narayanan, A., Voss, C., Yamaguchi, F., & Greenstadt, R. (2015). De-anonymizing programmers via code stylometry. In *24th USENIX Security Symposium (USENIX Security)*, Washington, DC. https://www.cs.drexel.edu/~ac993/papers/caliskan_deanonymizing.pdf

[21] Rosenblum, N., Zhu, X., & Miller, B. (2011). Who wrote this code? Identifying the authors of program binaries. *Computer Security–ESORICS, 2011*, 172–189. https://ftp.cs.wisc.edu/paradyn/papers/Rosenblum11Authorship.pdf

[22] Caliskan-Islam, A., Yamaguchi, F., Dauber, E., Harang, R., Rieck, K., Greenstadt, R., & Narayanan, A. (2015). When coding style survives compilation: De-anonymizing programmers from executable binaries. *arXiv preprint arXiv:1512.08546*. https://arxiv.org/pdf/1512.08546.pdf

[23] Bostrom, N. (2003). Are you living in a computer simulation? *Philosophical Quarterly, 53*(211), 243–255. https://ora.ox.ac.uk/objects/uuid:44c386c4-5d9e-4ecf-a47c-9631a2a59747/datastreams/ATTACHMENT01

[24] Hsu, S., & Zee, A. (2006). Message in the Sky. *Modern Physics Letters A, 21*(19), 1495–1500. https://arxiv.org/pdf/physics/0510102.pdf

[25] Beane, S. R., Davoudi, Z., & Savage, M. J. (2014). Constraints on the Universe as a Numerical Simulation. *The European Physical Journal A, 50*(9), 148. https://arxiv.org/pdf/1210.1847.pdf

[26] Hutchison, C. A., Chuang, R. Y., Noskov, V. N., Assad-Garcia, N., Deerinck, T. J., Ellisman, M. H., ... Pelletier, J. F. (2016). Design and synthesis of a minimal bacterial genome. *Science, 351*(6280), aad6253. https://pdfs.semanticscholar.org/9c1c/932fca27afa2ada8c1653ce9d22500e1abe6.pdf

[27] Katzenbeisser, S., & Petitcolas, F. (2000). *Information hiding techniques for steganography and digital watermarking*. Artech house.

[28] Beck, M. B., Rouchka, E. C., & Yampolskiy, R. V. (2012, October). Finding data in DNA: computer forensic investigations of living organisms. In *International Conference on Digital Forensics and Cyber Crime* (pp. 204–219). Springer, Berlin, Heidelberg. https://pdfs.semanticscholar.org/c8c3/fe9bb61cf44b29a1bdc1e40cdb5a6894c978.pdf

[29] Peirce, C.S. (1958). *Collected Writings* (8 Vols). (Ed. Charles Hartshorne, Paul Weiss & Arthur W. Burks). Cambridge, MA: Harvard University Press.

[30] de Saussure, Ferdinand (1983). *Course in General Linguistics* (trans. Roy Harris). London: Duckworth.

[31] Reed, M. L. (2000). Exosemiotics: an inter-disciplinary approach. *Acta Astronautica, 46*(10), 719–723.

[32] Freudenthal, H. (1960). *Lincos, Design of a Language for Cosmic Intercourse*. Amsterdam: North-Holland.

[33] Bailey, I. E., Backes, A., Walsh, P. T., Morgan, K. V., Meddle, S. L., & Healy, S. D. (2015). Image analysis of weaverbird nests reveals signature weave textures. *Royal Society Open Science, 2*(6), 150074. https://rsos.royalsocietypublishing.org/content/2/6/150074

[34] Hansell, M. H. (2005). *Animal Architecture*. Oxford: Oxford University Press, pp 321.

[35] Villaverde, A. F., & Banga, J. R. (2014). Reverse engineering and identification in systems biology: Strategies, perspectives and challenges. *Journal of the Royal Society Interface, 11*(91), 20130505. https://www.ncbi.nlm.nih.gov/pmc/articles/PMC3869153/

[36] Bostrom, N. (2012). The Superintelligent Will: Motivation and Instrumental Rationality in Advanced Artificial Agents. In V. C. Müller (Ed.), *Theory and Philosophy of AI*. Special issue, Minds and Machines, 22(2), 71–85.

[37] Omohundro, S. M. (2007). *The Nature of Self-Improving Artificial Intelligence*. Paper presented at *Singularity Summit 2007*, San Francisco, CA.

[38] Omohundro, S. M. (2008). The Basic AI Drives. In P. Wang, B. Goertzel & S. Franklin (Eds.), *Proceedings of the First AGI Conference, Frontiers in Artificial Intelligence and Applications, Volume 171*. Amsterdam: IOS Press, 483–492.

[39] Tegmark, M. (2017). *Life 3.0 Being Human in the Age of Artificial Intelligence*. New York: Knopf.

7

Preservation of Personal Identity: A Survey of Technological and Philosophical Scenarios

Soenke Ziesche and Roman V. Yampolskiy

7.1 Introduction

7.1.1 Personal Identity

Personal identity is the topic that deals with the question what is it that defines the continuity of a person over time. This was initially mostly discussed in philosophy, and three main approaches can be distinguished, the first two of which have in the meantime been pretty much excluded by science:

- Persistence through an extra-physical soul
- Bodily continuity
- Psychological continuity

Persistence through an extra-physical soul has been refuted, since science suggests that cognition is exclusively physical, which does not leave any space for the existence of a soul. And bodily persistence has been refuted, since there is no continuity of the human body as the atoms and molecules of our body, including our brain, are replaced regularly [1].

7.1.2 Psychological Continuity

Therefore, psychological continuity appears to be the scientifically supported foundation of personal identity. This means a mental relation is necessary to remain the same person, which includes, e.g., memories, beliefs and preferences over time, while neither a soul nor bodily persistence is required. On the philosophy side, this view can be traced back to Locke [2], while Parfit can be considered as one of the most influential and foresighted contemporary works in this field [3]. Parfit argued that the preservation of a so-called Relation R, which comprises psychological continuity, is the most important indicator of the continuous survival of a mind.

DOI: 10.1201/9781003565659-8

While the focus is here on human minds [e.g., 4], we can more generally define personal identity as a relation between two instantiations of minds, which is true when there is psychological continuity between these instantiations.

The acceptance of the psychological continuity approach was an important step to ponder the below introduced transfer of minds, since a transfer can be ruled out if the personal identity was linked to the body, let alone to an unfathomable soul.

7.1.3 Substrate-independence

Yet, in addition, the assumption of another concept is necessary, which is substrate-independence. As, e.g., Koene [5] or Tegmark [6] argue substrate-independence means that minds can be implemented on different physical substrates. What matters is that the same computational processes are executed. What does not matter is that these computations are implemented in carbon-based biological neural networks. This or very similar views have been referred to also by other terms, e.g., functionalism, computationalism [7] or patternism [1].

7.1.4 Transfer of Minds and Personal Identities

The topic of personal identity has been discussed for a long time without a specific field for application, but the discussion has gained only recently new relevance as well as dimensions when possibilities were considered to transfer human minds actively or passively to different substrates.

Since there is no evidence that immortality can be achieved with our biological bodies, scenarios have been contemplated, which would involve partly or fully transferring minds to other substrates. While these processes would mostly be controlled by the affected mind, philosophical reasoning (simulation) and scientific theories (multiverse) have come up with scenarios, in which minds could be transferred without the affected person having any impact on the process.

If we consider immortality as desirable or if we consider at least the possibility of immortality relevant as a research question and at the same time conclude that immortality in our biological bodies is impossible, all these concepts are only worth examining if they preserve the personal identity. This is because it is considered futile to transfer a mind, which is thereafter not anymore "the same" in the sense of personal identity.

While initially the criterion of psychological continuity for personal identity was only considered over time and thus rather straightforward, potential transfers of minds gave new dimensions to the question of sameness or similarity of the mind before and the mind after the transfer. In philosophy, Parfit described this first through thought experiments with a so-called teletransporter [3].

The main topic of this chapter is to survey what scenarios have been con-templated to transfer minds and how the chances are considered to preserve the personal identity of the affected mind.

7.2 Scenarios

In the following sub-sections, we introduce scenarios, which have in com-mon that a mind undergoes a transition. We survey for each scenario whether the transition is to be considered to preserve the personal identity of the affected mind. While the required technology is mostly work in progress and the philosophical simulation as well as the physical multiverse scenarios are speculative, the selection criteria for us were that there is no scientific evidence that these scenarios are not feasible.

As illustrated in Figure 7.1, the sub-section enhancements and upload are potentially for living minds. As various enhancements are likely to be ear-lier available than uploading technology, living enhanced minds could be uploaded once feasible. Cryonics and digital immortality rather function as

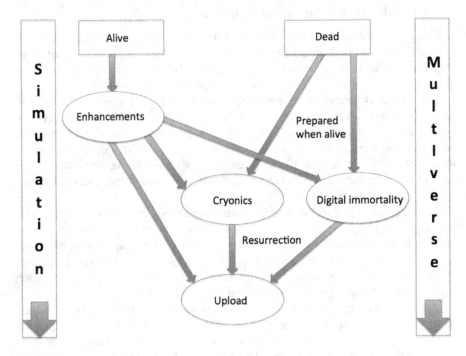

Figure 7.1 Overview of scenarios.

a means to an end and are considered as tools for (maybe already enhanced) minds whose embodiment expires before technology for upload exists to still enable an upload at a later stage, which would then involve resurrection. As we will explain simulation and multiverse could be epiphenomena of all these scenarios.

7.2.1 Enhancements

Bostrom defines a being as posthuman[1] if its capacities regarding health span, cognition and emotion vastly exceed current human beings [8]. In this scenario, we look only at posthuman cyborgs, which still have biological parts but are enhanced, e.g., with a brain–computer interface, as opposed to posthuman uploads, which will be introduced below.

Bostrom discusses whether these transformations would preserve personal identity. As far as an improved health span is concerned, which especially would result in a longer life, there appears to be no reason why such a healthier person would lose her or his identity. This assertion becomes even stronger if coupled with the concept that increased cognitive capacities imply especially better if not perfect memory capacities. Such a memory would secure psychological continuity even over long period of times and lifespans. Bostrom notices that the case is not that clear when it comes to radical cognitive or emotional enhancements and proposes the following six conditions:

- The changes are in the form of addition of new capacities or enhancement of old ones, without sacrifice of preexisting capacities.
- The changes are implemented gradually over an extended period of time.
- Each step of the transformation process is freely and competently chosen by the subject.
- The new capacities do not prevent the preexisting capacities from being periodically exercised.
- The subject retains her old memories and many of her basic desires and dispositions.
- The subject retains many of her old personal relationships and social connections; the transformation fits into the life narrative and self-conception of the subject.

Also, e.g., DiGrazia argues that enhancements are likely to preserve personal identity [11], while Schneider claims that there is not sufficient evidence for this [12]. Walker asserts that radical enhancements will not preserve personal identity and advocates like Bostrom for gradual changes [13].

A special case for cognitive enhancement is the so-called metaverse, which comprises virtual and augmented reality through ever improving devices, see, e.g., Vita-More [14]. While it is possible and often even intended to see "the world through the eyes of someone else" in the metaverse [e.g., 15], it is not believed that the personal identity of the mind experiencing the metaverse is lastingly affected.

7.2.2 Cryonics

Cryonics is a technique to preserve people soon after their death in liquid, very low-temperature nitrogen. The motivation is that the minds of these people can be resurrected if required technologies arise in the future. It is hoped that the preservation prevents damage and decay, but this is unproven in the long run.

Yet, if the preservation of the body works, advocates of cryonics, e.g., More [16] or Bostrom [9], believe that also the personal identity of the concerned mind can be preserved. Merkle introduces the concept of "information-theoretic death", which is the state of a mind when the necessary information to represent its personal identity cannot be recovered anymore [17]. Merkle claims that cryopreserved bodies, if correctly handled, have not experienced information-theoretic death.

Cryonics is offered by commercial services,[2] and also very recently, an alternative procedure for the preservation of a brain called vitrifixation has been introduced.[3]

As mentioned, cryonics is seen as a means to an end, which would ideally be completed by either resurrection within the biological body, which requires hypothetical nanomedicine, or upload via brain scanning. While the latter scenario is described below, the former is not considered here, since no transfer of the mind takes place.

7.2.3 Digital Immortality

Based on the central assumption that minds can be described by computational processes, efforts have been made to digitise the necessary data, i.e., the memories of human minds' lives. Such data have been captured since the invention of writing, but only recently:

- Digitisation became possible.
- More and more data are captured a) passively as exhaust data, e.g., through the use of devices, and sensor data, e.g., through wearables, and b) actively as produced digital content.
- In addition to other reasons for capturing these data, e.g., health monitoring, preservation of personal identity became a motivation.

Movements in this regard are called lifelogging and quantified self, and some pioneers were independently Bell [18], Wolfram [19] and Kurzweil [20]. Harari presents an even more comprehensive view and calls it dataism [21]. Especially social media data have triggered further projects, e.g., to create a digital avatar based on the digital content produced by a mind during embodiment.[4]

Just like cryonics, digital immortality is an attempt to potentially enable the resurrection of minds whose embodiments end before uploads are technically feasible. Turchin who works extensively on this topic considers digital immortality as "plan C" in this regard with life extension being plan A and cryonics plan B, also noting that digital immortality-based resurrection is significantly less expensive than cryonics [22]. Moreover, even those who can afford and have signed up for cryonics may die in such a way that they do not reach the cryonics facility in time or that the brain is damaged in such a way that it cannot be preserved.

Despite these efforts, challenges become obvious:

- Sufficiency of data: The amount of data collected is far less than the amount of memories and experiences stored in a human brain.[5] While it is assumed that it is possible to describe the human mind through data, current technologies are not sophisticated enough to capture all these data and can only be seen as a first step towards this goal. Turchin proposes a detailed methodology for this purpose, which focuses on active information-extraction efforts instead of passive recording [22]. Also, Bainbridge provides an approach for detailed personality capture and developed detailed questionnaires for this purpose [23]. Chalmers highlights the importance of brain scans and other medical and genetic data to be included in the records [24].

- Format of data: Since it cannot be assumed that whichever mind may carry out the upload reads and understands data that are stored in the way we are usually storing them, Wolfram outlines that "communication requires a certain sharing of cultural context" and attempts to tackle this challenge [25]. The format probably also plays a role for the later stage of resurrection, as given below.

- Preservation of data: As it is unknown when uploads will be technically feasible, the data should be preserved on a durable carrier in a secure location preventing decay and destruction. Turchin describes a strategy for this challenge [22].

- Accomplishing the resurrection: Even if all necessary data of a particular mind are available in digital format, we also have to know how to resurrect the relevant computations. Linked to this challenge is probably the fact that lifelogging data are usually not stored as neural networks as they are in the human brain. Vague proposals have been presented independently by Chalmers [24], Turchin [22] and Rothblatt [26]: Chalmers and Turchin state a superintelligent as well as benevolent AI in the future may be capable as well as motivated to resurrect minds based on the records prepared for digital immortality. In other words, the effort to collect and preserve all these data is for the non-zero probability that such an AI will appear, potentially through research on AI safety, which aims to ensure, somehow if possible at all, the above features of a superintelligent AI.[6] Rothblatt uses the concepts of "mindfile" for the digital immortality data and "mindware" for the required resurrection mechanism. Rothblatt cofounded a so-called Terasem Movement, which offers a service[7] to capture the mindfile, while the mindware does not exist yet.

7.2.4 Upload

As indicated before and illustrated in Figure 7.1, upload is seen by many as most durable scenario to preserve minds. Upload refers to the transfer of human minds to other physical substrates, e.g., a computer, due to the fragility of biological bodies, which should be feasible in theory after accepting the substrate-independence of minds. The details how to potentially realise such a transfer are described, e.g., by Koene [27], Sandberg and Bostrom, who presented a roadmap for whole brain emulation [28], or Wiley, who provides a taxonomy of potential mind-uploading procedures [29].

We can distinguish at least four variations of mind uploading as illustrated in Table 7.1 and described below.[8]

1. In this scenario, the upload would have the same perceptions of the real world as before through sensors, which have the same wavelengths as the human sensory system.

2. In this scenario, the upload's perceptions of the real world would be enhanced compared with humans, which should be technically feasible at a stage when uploads are technically feasible. Examples would be that these uploads could perceive extended frequencies, e.g., infrared radiation or ultrasound. This could be combined with further enhancements, e.g., a perfect memory.

TABLE 7.1

Upload Variations

	Real World	Simulation
Non-enhanced	1	3
Enhanced	2	4

3. In this scenario, the upload would have perceptions of a simulated world yet with the same wavelengths as the human sensory system.

4. In this scenario, the upload would have perceptions of a simulated world yet enhanced in most versatile ways, as Loosemore illustrates as follows: "Your choice of activities for the day might include: becoming a tiger and going off to the jungle for some animal sex; changing into a body that can swim in the atmosphere of Jupiter" [30].

The scenarios 1 and 3 may only be intermediate stages, while 2 and 4 are likely to be more attractive once feasible. Another option may be to switch between the variations, e.g., an uploaded mind may experience the real world and has the opportunity to spend time in a simulation, compared with the experience in a metaverse as described above.

Since hopes are placed in uploading, also the question of the preservation of personal identity has been discussed extensively. Sandberg and Bostrom believe their approach would preserve personal identity even if the brain activity temporarily ceased as cases of coma as well as hypothermia patients have shown [28]. Walker approaches the topic by using the type/token distinction and argues that uploading would sacrifice the token identity, which "is not inconsiderable" but preserve the type identity, which "will be in better shape than the original token", e.g., regarding immortality and enhancement [31].

Since we assumed substrate-independence, there is not so much concern whether a mind can be implemented on a computer, but there are rather debates about the potential loss of personal identity during the uploading procedure:

According to Chalmers, the chances to preserve personal identity depends on the method of the upload, of which he distinguishes between destructive uploading, non-destructive uploading, gradual uploading and reconstructive uploading. He is "reasonably confident that gradual uploading is a form of survival" [24].

Hauskeller argues that there is no evidence that this will be possible and points out that an "argument from graduality", beginning from the Ship of Theseus, is "always fallacious because it denies the reality of change" [32]. Also, Corabi and Schneider [33] disagree with Chalmers and are pessimistic that destructive, non-destructive or even gradual uploading can preserve personal identity.

Simulations have been mentioned already in scenarios 3 and 4, but in the following section, the focus is more on the controlling role of the simulating mind.

7.2.5 Simulation and Resurrection in Different Simulations

This section is based on Bostrom's simulation argument and contains, on the one hand, the most speculative and hardest to prove scenarios, and on the other hand, due to its potential the most diverse scenarios [34]. The main part focuses on scenarios assuming that we are currently in a simulation already, for which the probabilities are elaborated by Bostrom. Simulation scenarios have been not much discussed regarding preservation of personal identity but have very interesting aspects as we outline here.

If we are currently in a simulation, we can conclude that 1) the data of our personal identity are somehow captured by software and 2) these data can be retrieved in another simulation. In fact, this scenario is the most certain one for preservation of personal identity, hence for immortality of minds, yet probably beyond our control and dependent entirely on the simulating mind.

It is worth noting that for this scenario, we do not have to assume substrate-independence of minds. It is sufficient that there is one substrate, used by the simulating mind, which is appropriate to run such a simulation (and which could be silicon-based, but it does not matter).

Apart from discussions about probabilities that we are in a simulation, we do not know anything about the motives of the simulating mind (which could even be itself within a simulation, i.e., there could be scenarios of nested simulations) [35]. Therefore, we shall assume according to Bostrom's orthogonality thesis that "more or less any level of intelligence could in principle be combined with more or less any final goal" [36]. This means regarding the transfer of simulated minds from one simulation to another that basically anything could be possible. The extreme scenarios for our universe would be, on the one hand, that every mind that was ever simulated on earth has been transferred to another simulation and resurrected there with full preservation of personal identity and, on the other hand, that not a single mind that was ever simulated on earth has ever been resurrected in another simulation.

In addition, countless other scenarios are imaginable such as the following:

- Some minds are resurrected in another simulation according to a reward system, which is unbeknown to us, while some minds are never resurrected according to a penalisation system, which is unbeknown to us as well.[9]

- Some minds are resurrected in another simulation according to criteria, which could be as random as that only minds that during their simulation calculated correctly at least once the sum of 793 and 826 will be resurrected.

- Some minds are resurrected in another simulation but without preservation of personal identity (although preservation of personal identity is technically feasible within simulations as we established above).

The last example brings us to the question, still assuming that we are in a simulation, does the fact that we do not have memories of any previous lives mean that we have not been simulated before or could we (or some of us) have been simulated before, but for reasons unbeknown to us the simulating mind resurrected us in this simulation without preservation of personal identity, i.e., such memories? Could this also mean that while currently we do not experience psychological-continuity with previous simulations that nevertheless the necessary data may have been archived by the simulating mind?

If this was affirmative, it may be possible that in yet another resurrection, we may get access to the memories of our previous personal identity again. And we could think of scenarios where we are resurrected in a potentially completely different simulation[10] (e.g., different gender, different species, different qualia, different physical laws and even immortal) where we remember the preceding simulation although it was completely different. And since the simulating mind may have saved our full personal identity, there may be resurrections, in which we remember not only this currently simulated life but also other ones before. Again, options are countless: The simulating mind, which archives the full history of all simulated minds, may or may not for reasons unbeknown to us make parts of that personal identity accessible and/or may even modify memories or include fake memories.

The idea that the simulating mind may have all data of all instantiations of the simulated minds could add another dimension to the concept of extended minds developed by Clark and Chalmers [37]. They claim that objects such as a notebook could function as a part of the mind, if it includes information that cannot be retrieved from the brain. Yet, in the scenario described here, the extended mind of the simulated mind would comprise parts of the simulating mind, since it holds an abundance of information not directly accessible by the simulated mind.

While we can conclude that if we are in a simulation in theory, our personal identity can be preserved if resurrected in a further simulation, we may have to re-define personal identity and also psychological-continuity for these diverse potential scenarios, which is obviously severely hampered by lacking evidence, but will nevertheless be elaborated on below.

Let us also look at the other scenario that we are currently not in a simulation. Then, there are two options: First, there will never be simulations. We do not have to look at this option. Second, there will be simulations in the future.[11] In this case, the question arises whether resurrection by preserving personal identity would be possible, after, e.g., measures of cryonics or digital immortality were applied until simulations became available. This in turn depends on the motivations and capacities of the future simulating mind.

Finally, it has to be highlighted that as illustrated in Figure 7.1, all other scenarios described here could occur within a simulation, which leads to another variety of (highly speculative) scenarios. Two extremes would be that maybe the simulating mind anyway resurrects every mind, and human minds actually do not have to worry about immortality and uploading, etc., or if human minds finally figure out how to successfully upload themselves, the simulating mind might turn the simulation off for reasons unbeknown to us.

7.2.6 Multiverse

Like the simulation scenario, the multiverse scenario can be seen as an epiphenomenon to the previous scenarios, both being characterised as hard to prove and perhaps uncontrollable by current human minds. The concept is based on quantum mechanics and its many-worlds interpretation by Everett [38].[12] In brief, this interpretation asserts that the wave function in quantum mechanics does not collapse, and as a consequence, the worlds branch off constantly to all possible futures, which then leads to the existence of an unimaginable amount of worlds in parallel, referred to as multiverse.

Tegmark and others infer from this many-worlds interpretation a thought experiment that for any life-threatening situation, there will always be a resulting world where the concerned mind survives, referred to as quantum immortality [40]. This leads again to a variety of scenarios, outlined by Turchin, who, e.g., describes that there may be immortality, yet in many instances combined with suffering [41]. In this context, Tegmark also describes a related thought experiment about attempted quantum suicide and its guaranteed survival in one branch, which could actually lead to a discussion about one's right to quantum suicide similar to the discussion about a right to euthanasia [40].

Since the focus of this chapter is on preservation of personal identity, the question is whether in these multiverse branches, where the mind survives, the personal identity of the mind is preserved too. For example, Tappenden [42] and Ćirković [43] claim that this is the case. And this is what human minds, if this many-worlds interpretation is correct, feel is the case for each particular branch, in which they live for an instant, thus which they observe, i.e., that they remain the same person despite constant unnoticeable branching.

7.3 Potential Features of Mind and Personal Identity Transfers

As we conclude the survey of scenarios, it became evident that some potential features of personal identity transfers are relevant, on which we elaborate below: Fission, fusion, resurrection and different cognition. Fission in particular and also fusion of minds have been discussed extensively but are not in the focus of this chapter, since both features are not likely to be desirable for a human mind striving for immortality. We shall in particular focus on different cognitive abilities, which has not been researched much before in relation to personal identity.

7.3.1 Fission

Since Parfit's original work [3], fission scenarios have been frequently debated, especially in the light of potentially upcoming technologies. If such technologies enable us to create replicas of a mind, the consequences for the personal identity have to be examined. For example, Walker describes it light-heartedly: "if there are a thousand replicas, then they will quickly have psychologically distinguishable properties. All thousand replicas will not fit in the same cab, e.g., and so will have different experiences leaving the replicating center" [31]. This means while the replicas still have psychological continuity with the original, they do not have psychological continuity between each other. Wiley argues that all replicas "are truly equal in the primacy of their claim to the identity of the original mind" [29].

If Everett's many-worlds interpretation was correct, then there would be constant fission taking place on an unimaginable scale. In general, if the data of our personal identity could be captured by software, as assumed for the upload and the simulation scenarios, fission becomes basically trivial by creating copies of these data. And unlike in the multiverse scenario there could be several copies of a mind in the same universe or simulation too.

7.3.2 Fusion

Scenarios have been suggested, which entail fusion of minds. Goertzel [44], Sotala and Valpola [45] as well as Ziesche and Yampolskiy [35] present overviews. Sotala and Valpola discuss the two options, either that the resulting mind keeps the personal identities of both original minds or that the resulting mind has an entirely new personal identity.

Since fusion of minds is, despite the above attempts, very hard to conceptualise, it is also hard to explore if in the above scenarios fusion of minds may be incorporated, let alone what consequences this has for the affected personal identities.

7.3.3 Resurrection

Another question is whether psychological continuity would be affected by break times or, to phrase it the other way round, can personal identity be preserved if the concerned mind is not continuously animated? This question is particularly relevant when possibilities for resurrection of minds, which have ceased to exist, are assessed. One could argue for a trivial solution that already sleeping constitutes a break time of psychological continuity, which, however, has no impact at all on the personal identity after the concerned person wakes up.

As described, for resurrection, there have been no specific proposals presented yet, except for "outsourcing" it to a future AI [e.g., 22]. Only if the simulation scenario was affirmative, which comprises so many potential variations, resurrection should also be feasible within it.

7.3.4 Different Cognitive Abilities

As we saw above, an epiphenomenon of the transfer of minds to other substrates may likely be that there is a difference between the capacities of the mind before and after the transfer, which may include different cognitive abilities. While transhumanists strive for enhancement of capacities, we look also at reduction of capacities, which may be relevant for the transfer between simulations with many more arbitrary parameters. The question, which appears not to have been researched much, is whether the change of the potential capacities has an impact on the preservation of the personal identity? Bostrom [8] and also Walker [13] emphasise that changes should be implemented gradually and slowly, which may not necessarily be controllable.

We aim to illustrate the issues through the following examples regarding psychological continuity:

7.3.4.1 Reduced Cognitive Abilities

Let us consider a mind, which is able, e.g., to perceive in one instantiation ultrasonic frequencies, but in a later instantiation, perhaps after having been transferred to another substrate, the mind cannot perceive these frequencies anymore, nor can any other mind in this environment perceive these frequencies. Can this mind have memories of its ultrasonic experiences? Let us further assume that ultrasonic sound played a crucial role for this mind. For example, a central purpose of existence for this mind may have been to compose some kind of ultrasonic music. Is there still psychological continuity between the instances of this mind if crucial perceptions are lost?

It may be claimed that this scenario could, e.g., be compared with Beethoven who went deaf, yet not only preserved his personal identity, but also must have had uninterrupted memories of sounds as he continued to successfully compose music. Yet this comparison is not entirely valid, since Beethoven remained in an environment where the vast majority of fellow minds was familiar with sounds.

We should note that this issue does not exist if it was the other way around, i.e., if the perception range increases from instance to instance, e.g., through transhuman features or potentially through uploading, see above.

7.3.4.2 Massively Enhanced Cognitive Abilities

Another argument is linked to the bias that we know only the range of perception we are experiencing in our current instantiation. We can imagine instantiations, which perceive a much higher amount of perceptions, e.g., due to massively enhanced cognition, much more complex environments and/or much longer embodiment. This may entail that the ratio of memories associated with our current instantiation becomes tiny in relation to the memories of the later, much more multifaceted instantiations. And this leads to the question whether there will remain psychological continuity with our current instantiation for a mind in such a multifaceted instantiation, notwithstanding a near perfect memory thanks to enhancement, because the current instantiation may fade as an unimpressive episode?

7.3.4.3 Contradictory Qualia

As, e.g., described by Yampolskiy [46] and also by Loosemore [30], further potentially possible qualia are hard to imagine from a human centric view, yet as motivated by Ziesche and Yampolskiy perhaps essential for survival in order to prevent boredom during very long lifespans [47]. Especially in different simulations, but possibly already through enhancements novel qualia may be experienced. Such qualia may contradict qualia from the current instantiations, which could be as simple as swapped perception of certain colours, yet potentially much more complex. The question arises if memories of the current instantiation would be then suppressed, since they are too confusing in light of the contradictory qualia, which may in turn also have an impact on the psychological continuity, thus the preservation of personal identity?

7.4 Conclusion

7.4.1 Immortality

We have shown that enhancements through brain–computer interfaces are considered as first step towards posthuman life-extension. Yet, immortality of a mind may only be ensured if entire uploads of the mind to a different substrate were possible. Since the required technologies are not yet available, in order to gain time, two means to this end have been envisaged: cryonics and digital immortality. However, they would both require resurrection, which has also not been undertaken by now. Furthermore, we presented a philosophical (simulation) and a scientific (multiverse) scenario, which have a non-zero probability. Both may ensure immortality, yet do we neither have evidence for it nor control.

7.4.2 Personal Identity

As motivated before, for human minds, immortality is likely not to be desirable if the personal identity is not preserved. In Table 7.2, we summarise the presented scenarios and potential features. As described above, due to lack of evidence for none of these scenarios, the preservation of personal identity is guaranteed. Likewise, it has not been proven that for any of these scenarios the personal identity will not be preserved. Nevertheless, it must not be argued that this is, until there is evidence, an idle discussion because of the immense relevance of the topic and the potential consequences.

We have used the psychological continuity approach to survey the views on personal identity in the presented scenarios. We have shown that especially for upload and simulation, if feasible or true, respectively, our cognition may be enhanced or changed in a way perhaps impossible to imagine from our human centric view. The question arises whether the psychological continuity approach is robust enough to incorporate such scenarios or whether this approach has to be extended?

7.4.3 Extended Approach

As outlined especially for the simulation scenario, it may be that each instantiation of a particular mind has different sets of memories, perceived qualia and mental states, the intersections of which may be empty or not. The overall personal identity of a mind would then be expressed by the union of these sets, to which the mind may or may not have access depending on the instantiation. Nevertheless, this union can be described for each mind (and the simulating mind may have access). We proposed that it might be promising to expand the extended mind approach from Clark and Chalmers [37] to define the personal identity of minds, which undergo massive changes of cognition. The union of all sets of memories, perceived qualia and mental states would correspond to the extended mind conceptualised by Clark and Chalmers. Set theory could be also applied to describe the personal identity of minds, which undergo fission or fusion.

Goertzel presents a method to measure if two minds should be considered versions of one another, yet without the groundwork that we proposed above [48]. Yampolskiy proposes for the same task a hypothetical variant of the Turing Test, according to which the original mind would ask queries to the mind in question, to which supposedly only the original mind would know answers [4]. Suitable questions would "relate to personal preferences, secrets (e.g. passwords) as well as recent dreams". While we proposed above that the union of these sets could express the personal identity of a mind, we

TABLE 7.2

Summary of Scenarios

	Preservation of Personal Identity	Process Controlled by Affected Mind	Fission Possible	Fusion Possible	Different Cognitive Abilities Possible	Resurrection Possible
Enhancement	Not disputed for smaller enhancements, disputed for complex enhancements	Yes (as far as decision to become enhanced is concerned)	No	No (maybe for brain–computer interfaces involving more than one mind?)	Yes	n/a
Cryonics	Means to an end, resurrection disputed	Yes (as far as decision to get cryonised is concerned)	n/a	n/a	n/a	Envisaged
Digital immortality	Means to an end, resurrection disputed	Yes (as far as data collection is concerned)	n/a	n/a	n/a	Envisaged
Mind uploading	Disputed	Yes (if alive), no (if resurrection)	Yes	?	Yes	Envisaged
Simulation	Too speculative to judge	No	Yes	?	Yes	Yes
Multiverse	Too speculative to judge	No	Yes	?	No	No

could use the intersection operation to examine the similarity between two minds in different instantiations. If we clone a mind, the copies will quickly start diverging but would certainly have, at least initially, a very high percentage of similarity with the original mind. This is expressed by a relatively high cardinality of the intersecting set of the sets of memories, perceived qualia and mental states of the original mind and its copies. Since the cardinality of this intersecting set gets inevitably smaller over time even for a conventional human mind notwithstanding all the above scenarios, one could even argue that persistence of personal identity is an illusion and we are simply change blind because of how gradually it happens in most cases.

This new approach to define the personal identity through means of set theory can be seen as a contribution to intellectology, a new field of study introduced by Yampolskiy with the intention to examine the overall space of minds not limited by a human centric view.

To summarise, in addition to surveying how chances of preservation of personal identity are currently assessed for potential scenarios to establish immortality of minds, we also presented the following novel contributions: if transfer of minds is feasible (and ideally even controllable), the abilities of the transferred mind may be extremely different in many dimensions, which may currently be underestimated due to a human centric view. Yet we highlighted that these possibilities have to be taken into consideration when debating the preservation of personal identity. Therefore, we elaborated on different cognitive abilities as well as on the enormous potential of the simulation scenario. In order to cope with these options and to still sufficiently specify the preservation of personal identity, we outlined how the psychological continuity approach may have to be expanded. This idea leaves various challenges for future work.

Notes

1 In this context, also the term transhumanism is used, which we will not need here, since it refers to an intermediary stage as well as the movement advocating for and researching posthuman capacities [9, 10].

2 See, e.g., http://alcor.org/.

3 See https://nectome.com/.

4 See https://eternime.breezy.hr/.

5 It would not be correct to state that the collected data constitute a subset of the memories and experiences stored in a human mind, since it is very likely that the collected data include some information, which the mind has forgotten or to which it has no access anymore. For example, video recordings of a body cam contain much more information than a mind ever stores in the long term, but also diary entries are unlikely to be memorised verbatim.

6 Another undesired but also non-zero probability scenario would be that a malevolent AI appears and resurrects these minds with the intention to conduct harmful experiments or simply to torture them, while minds whose data were not preserved are spared from this ordeal.
7 See https://www.lifenaut.com/.
8 As described below we do not know if we are currently in a real world or if this is already a simulation. Here we refer with "real world" to where we live currently.
9 We do not go into detail here, but other simulations, in which these minds may be resurrected, could be entirely different from our universe and beyond imagination for us. We only note the caveat that mind transfer and resurrection per se may not always be desirable as the next simulation could have a much worse "quality of life".
10 While Bostrom focuses on ancestor-simulations [34], we are looking at broader options.
11 Note Bostrom's probability assessment of these options: "Unless we are now living in a simulation, our descendants will almost certainly never run an ancestor-simulation" [34, p. 14].
12 It has to be highlighted that there is more than one many-worlds interpretation and that these many-worlds interpretations also have many opponents [e.g., 39].

References

[1] Kurzweil, R. (2005). *The Singularity Is Near: When Humans Transcend Biology.* Viking.
[2] Milton, J.R. (1694). John Locke: An Essay concerning Human Understanding. In *Central Works of Philosophy v2* (pp. 115–136). Routledge.
[3] Parfit, D. (1984). *Reasons and Persons.* OUP Oxford.
[4] Yampolskiy, R.V. (2015). *Artificial Superintelligence: a Futuristic Approach.* Chapman and Hall/CRC Press (Taylor & Francis Group).
[5] Koene, R. A. (2012). Embracing competitive balance: The case for substrate-independent minds and whole brain emulation. *Singularity Hypotheses.* Springer, Berlin, Heidelberg, 241–267.
[6] Tegmark, M. (2017). Substrate-Independence. In J. Brockman (Ed.), *This Idea Is Brilliant: Lost, Overlooked, and Underappreciated Scientific Concepts Everyone Should Know.* Harper Perennial. https://www.edge.org/response-detail/27126
[7] Putnam, H. (1960). *Minds and Machines.* Dimensions of Mind/New York University Press.
[8] Bostrom, N. (2008). Why I want to be a posthuman when I grow up. *Medical enhancement and posthumanity.* Springer, Dordrecht, 107–136. https://nickbostrom.com/posthuman.pdf
[9] Bostrom, N. (2003). *The Transhumanist FAQ: v 2.1.* World Transhumanist Association. https://nickbostrom.com/views/transhumanist.pdf

[10] More, M. (1990). Transhumanism: Towards a futurist philosophy. *Extropy*, 6, 6–12. http://fennetic.net/irc/extropy/ext6.pdf

[11] DeGrazia, D. (2005). Enhancement technologies and human identity. *Journal of Medicine and Philosophy*, 30(3), 261–283. https://www.tandfonline.com/doi/full/10.1080/03605310590960166

[12] Schneider, S. (2008). Future minds: Transhumanism, cognitive enhancement and the nature of persons. *Neuroethics Publications*, 37. https://repository.upenn.edu/cgi/viewcontent.cgi?article=1037&context=neuroethics_pubs

[13] Walker, M. (2008). Cognitive enhancement and the identity objection. *Journal of Evolution and Technology*, 18(1), 108–115. https://jetpress.org/v18/walker.htm

[14] Vita-More, N. (2010), Epoch of plasticity: The metaverse as a vehicle for cognitive enhancement. *Metaverse Creativity*, 1, 1, 69–80. https://pdfs.semanticscholar.org/61af/8be2b6cab162b0c2257f080d44bffd7a7cae.pdf

[15] Harth, J. (2018). Being There, Being Someone Else: Leisure and Identity in the Age of Virtual Reality. In *Global Leisure and the Struggle for a Better World*. Palgrave Macmillan, Cham, 141–159.

[16] More, M. (1995). *The Diachronic Self: Identity, Continuity, Transformation.* University of Southern California.

[17] Merkle, R. C. (1992). The technical feasibility of cryonics. *Medical Hypotheses*, 39(1), 6–16.

[18] Bell, C. G., Gemmell, J., & Gates, B. (2009). *Total Recall: How the e-memory Revolution Will Change Everything.* New York: Dutton.

[19] Wolfram, S. (2012). The Personal Analytics of My Life. http://blog.stephenwolfram.com/2012/03/the-personal-analytics-of-my-life/

[20] Kushner, D. (2009). When Man and Machine merge. Interview with Ray Kurzweil. *Rolling Stone Magazine*.

[21] Harari, Y. N. (2016). *Homo Deus: A Brief History of Tomorrow.* Random House.

[22] Turchin, A. (2018). Digital Immortality: Theory and Protocol for Indirect Mind Uploading. https://philpapers.org/rec/TURDIT

[23] Bainbridge, W. S. (2013). *Personality Capture and Emulation.* Springer Science & Business Media.

[24] Chalmers, D. (2014). Uploading: A philosophical analysis. In R. Blackford & D. Broderick. (Eds.), *Intelligence Unbound: The Future of Uploaded and Machine Minds*. Chichester: John Wiley & Sons, 102–118. http://consc.net/papers/uploading.pdf

[25] Wolfram, S. (2018). How to design beacons for humanity's afterlife. *WIRED.* https://www.wired.com/story/how-to-design-beacons-for-humanitys-afterlife/

[26] Rothblatt, M. (2012). The Terasem mind uploading experiment. *International Journal of Machine Consciousness*, 4(1), 141–158. https://www.cyberev.org/martine.pdf

[27] Koene, R. A. (2014). Feasible mind uploading. In R. Blackford & D. Broderick. (Eds.), *Intelligence Unbound: The Future of Uploaded and Machine Minds.* Chichester: John Wiley & Sons, 90–101.

[28] Sandberg, A., & Bostrom, N. (2008). *Whole Brain Emulation: A Roadmap Technical Report.* Future of Humanity Institute. https://www.fhi.ox.ac.uk/brain-emulation-roadmap-report.pdf

[29] Wiley, K. (2014). *A Taxonomy and Metaphysics of Mind-uploading.* Humanity+ Press and Alautun Press.

[30] Loosemore, R.P.W. (2014). Qualia Surfing. In R. Blackford & D. Broderick. (Eds.), *Intelligence Unbound: The Future of Uploaded and Machine Minds*. Chichester: John Wiley & Sons, 231–239. http://www.richardloosemore.com/wp-content/uploads/2017/05/2014b_QualiaSurfing_rpwl.pdf

[31] Walker, M. (2011). Personal identity and uploading. *Journal of Evolution and Technology*, 22(1), 37–51. https://jetpress.org/v22/walker.htm

[32] Hauskeller, M. (2012). My brain, my mind, and I: some philosophical assumptions of mind-uploading. *International Journal of Machine Consciousness*, 4(1), 187–200. http://transhumanism102.weebly.com/uploads/1/8/3/4/18348243/mind_uploading_source_4.pdf

[33] Corabi, J., & Schneider, S. (2014). If You Upload, Will You Survive?. In R. Blackford & D. Broderick. (Eds.), *Intelligence Unbound: The Future of Uploaded and Machine Minds*. Chichester: John Wiley & Sons, 131–145.

[34] Bostrom, N. (2003). Are we living in a computer simulation?. *The Philosophical Quarterly*, 53(211), 243–255. https://www.simulation-argument.com/simulation.pdf

[35] Ziesche, S. & Yampolskiy, R. V. (2020). Towards the Mathematics of Intelligence. In *The Age of Artificial Intelligence: An Exploration*, ed. S.S. Gouveia, 3–13. Wilmington: Vernon Press.

[36] Bostrom, N. (2012). The superintelligent will: Motivation and instrumental rationality in advanced artificial agents. *Minds and Machines*, 22(2), 71–85. https://nickbostrom.com/superintelligentwill.pdf

[37] Clark, A., & Chalmers, D. (1998). The extended mind. *Analysis*, 58(1), 7–19. http://cogprints.org/320/1/extended.html

[38] Everett III, H. (1957). "Relative state" formulation of quantum mechanics. *Reviews of Modern Physics*, 29(3), 454.

[39] Kent, A. (1990). Against many-worlds interpretations. *International Journal of Modern Physics A*, 5(09), 1745–1762. https://pdfs.semanticscholar.org/b018/30bb0ec078ff58f143a3ff18961b92ca95ec.pdf

[40] Tegmark, M. (1998). The interpretation of quantum mechanics: Many worlds or many words? *Fortschritte der Physik: Progress of Physics*, 46(6–8), 855–862. https://arxiv.org/pdf/quant-ph/9709032.pdf

[41] Turchin, A. (submitted). Forever and Again: Necessary Conditions for "Quantum Immortality" and its Practical Implications. https://philpapers.org/archive/TURFAA-3.docx

[42] Tappenden, P. (2000). Identity and probability in Everett's multiverse. *The British Journal for the Philosophy of Science*, 51(1), 99–114.

[43] Ćirković, M. M. (2006). Is quantum suicide painless? On an apparent violation of the principal principle. *Foundations of science*, 11(3), 287–296. https://arxiv.org/pdf/quant-ph/0412147.pdf

[44] Goertzel, B. (2003). Mindplexes - The Potential Emergence of Multiple Levels of Focused Consciousness in Communities of AI's and Humans. Dynamical Psychology.

[45] Sotala, K., & Valpola, H. (2012). Coalescing minds: Brain uploading-related group mind scenarios. *International Journal of Machine Consciousness*, 4(1), 293–312. https://intelligence.org/files/CoalescingMinds.pdf

[46] Yampolskiy, R. V. (2017). Detecting Qualia in Natural and Artificial Agents. arXiv preprint arXiv:1712.04020.

[47] Ziesche, S., & Yampolskiy, R. V. (2017). High Performance Computing of Possible Minds. *International Journal of Grid and High Performance Computing (IJGHPC)*, 9(1), 37–47.

[48] Goertzel, B. (2012). When should two minds be considered versions of one another? *International Journal of Machine Consciousness*, 4(1), 177–185. https://pdfs.semanticscholar.org/ccfb/0b844be4d306f164320636bed8653b699ec1.pdf

8

The Problem of AI Identity

Soenke Ziesche and Roman V. Yampolskiy

8.1 Introduction

Among various identity problems, such as the Ship of Theseus [1, 2], a particular one, which has been examined at length, is the problem of personal identity, which addresses the question of what it is that defines the continuity of a person over time [3]. Although no final consensus for this problem has been reached, we propose to commence the endeavour towards AI identity by looking at personal identity. Humans, at least during some parts of their life span, and AIs have in common that they reach certain states over time by processing information while applying intelligence, and this ability appears to be a critical component according to some approaches towards the problem of personal identity, which may be also harnessed to characterise AI identity over time as will be outlined below.

Whereas most humans over large parts of their life span also have the feature of being sentient, this is not clear for AIs, and actually by many not even considered at all. If (some) AIs were to be sentient, they would be, just like humans, a subset of the universe of minds [4]. It has been argued that there is a non-zero probability for the existence of such sentient AIs; thus, their welfare deserves critical consideration, which includes the prevention of suffering and of deletion of such minds [5, 6]. While we look here at both non-sentient and potentially sentient AIs, the question of identity is especially relevant for sentient AIs as a prerequisite to address their individual welfare.

This chapter is structured as follows: first, we motivate the AI identity problem by outlining several fields of application. This is followed by the first proposed approach to tackle the problem, which includes an introduction to the personal identity problem. Thereafter, the second proposed approach, based on multi-factor authentication, is outlined. We then examine a variety of scenarios. In the end, a summary and an outlook for future work is presented.

DOI: 10.1201/9781003565659-9

8.2 Motivation

Why is it critical to explore the AI identity problem? We have identified several areas of interest, which also shows that we aim to look at the problem from a wide lens, which includes philosophical aspects, particularly the following aspects:

1. It is relevant for legal matters to determine the criteria when an AI remains the same because only then ownerships [7, 8] and patents [9] for AIs may be claimed and tracked over time.[1]

2. A different dimension compared with ownership of AI would be personhood of AI, which is a topic of debates [10] and which, if granted, would also assign accountability to AIs, e.g., related to business relationships with them. This in turn would necessitate clarity about their identity over time.

3. Yet a further dimension would be if sentient AIs exist or come into existence. The identity of such AIs over time is relevant to study for various reasons, not only to ensure their welfare, as indicated above [6] but, e.g., also for future scenarios of relationships such as friendship or even marriage with sentient AIs.

4. Work is ongoing towards brain–machine interfaces involving AI [11], and future scenarios may envisage uploaded human minds merged with AIs. Both cases also require clarity about the identity of the involved AI over time.

5. AIs may also collaborate as multi-agent systems, which may be for an observer undistinguishable from a singleton AI [12]. Insights to the AI identity problem would help to figure out whether it is a society of agents or a single agent we are dealing with.

6. Finally, this research may help solving the problem of personal identity. Many debates about the problem of personal identity get complicated by components, such as consciousness, which are beyond the realm of contemporary science. In contrast, AI systems, i.e., computational processes, are more open to methods of contemporary science.

In summary, while the chapter focuses on AI identity, it is also envisaged that this endeavour may inform our understanding of humans. Therefore, the last point problem of personal identity in particular is picked up again in the summary below.

8.3 First Approach

Given the possible similarities between the AI identity problem and the personal identity problem and the fact that the latter one has been researched for centuries, we introduce it here and examine what aspects can be transferred to the AI identity problem.

8.3.1 The Personal Identity Problem

As it is, e.g., outlined in an overview about the preservation of personal identity by Yampolskiy and Ziesche [13], various views on personal identity have been developed, two of which are briefly introduced here. The animalism approach asserts that humans are a proper subset of all animals or organisms. Conversely, there are organisms, who are not persons. According to this view, being a person is merely a temporary property of us, which implies that humans are both body and soul [14]. Another contrasting view is called psychological continuity, which can be traced back to Locke [15]. This approach can be characterised as follows: There is a relation, R, of psychological continuity such that a person x at a time t is the same person as a person y at a time t^*, if and only if x at time t bears that relation R to y at t^* [3]. As mentioned, no final consensus for this problem has been reached, yet one of the more recent approaches by Shoemaker appears applicable to the AI identity problem [16], as will be outlined. He distinguishes between psychological connectedness and psychological continuity. A person is psychologically connected with a person in the past if she or he is now in psychological states *because of* psychological states she or he was in in the past. In other words, there is a causal relation between these psychological states. According to Shoemaker, there is personal identity between persons at different points in time when there is psychological continuity, and there is psychological continuity between persons at different points in time when the psychological states at a later point relate to those at an earlier point by a chain of psychological connections.

The topic of personal identity has gained recently new relevance, since potential possibilities are discussed to transfer human minds to different substrates [e.g., 17], e.g., so-called uploading of human minds to a computer as a hypothetically more durable scenario to preserve human minds [18]. For such, yet still theoretical scenarios, some sub-scenarios are distinguished by Yampolskiy and Ziesche [13], such as fission, fusion and resurrection of human minds. Fission, which is the separation of a single entity, e.g., a person or an AI, into two or more parts, is examined by Parfit in detail [19] and is for humans rather a topic of thought experiments, yet for AIs due to their copyability simple to realise. To examine whether after fission the persons have the same personal identity as the original person, a concept called "multiple-occupancy view" has been developed [20]. According to this approach, the post-fission persons existed already prior to fission. This can be illustrated by a railroad track, which forks, potentially multiple times. While there is one track only, it figuratively overlaps all the branches to come. The philosophical position that one and the same physical object can be a proper part of two or more distinct objects is called "four-dimensionalism" [21].

8.3.2 Proposed Definition

We suggest exploring whether Shoemaker's notions of psychological connectedness and psychological continuity [16] can be adapted towards AI identity.

This leads us to propose the following definition: if the state of a current AI has been caused by a state of one specific AI in the past, then there is a chain of connections between these AIs, then there is continuity between these AIs and then there is identity between these AIs at different points in time.

For this definition, we define intelligence as "an agent's ability to achieve goals in a wide range of environments" [22]. If the "agent" in this definition is a human or a non-human animal, it is conventional intelligence, while it is AI if the agent is a machine. We also refer to the prevalent computer hardware substrate of AIs, which is based on a bunch of electric circuits and operating on two different voltage levels. Such a system is discrete, and the set of states it can occupy is well defined, countable and finite. This applies also to AIs, which operate on nowadays prevalent neural networks. While the neural networks use continuous functions, their implementation is discrete, since the underlying hardware is a discrete computer.

It is also important to clarify what is not required to ensure identity: identity between two AIs at different points in time does not require that the two AIs remain exactly the same over time, since we have seen that this is not a requirement for personal identity either. Here are two examples: first, there can be still AI identity over time when the AI has (significantly) increased its knowledge and performance over time; second, there can be still AI identity over time when the AI takes a "treacherous turn" [23].

8.3.3 Concept of Computational Irreducibility

For the verification of this definition, we introduce a method, which harnesses again the computer hardware substrate of AIs, i.e., the fact that computational processes are easier to formalise than neuronal ones. We propose to apply the concept of computational irreducibility [24] and argue that there is identity between two AIs if the latest one cannot be produced by any short-cut but has to be computed from the original one. This can be further refined by enumerating the space of AIs as it has been done, e.g., for the space of minds by Yampolskiy [4]. If we assign an integer to any AI, we can map it to states of specific cellular automata. To do so, we take the integer representing a specific AI and convert it to the binary code with each zero and one representing a state of a cell in a cellular automaton (off or on). Running this cellular automaton shows all the other cellular automata states it is connected to in a computationally irreducible manner. Taking the state of the cellular automata in terms of on and off cells and converting it back to binary code and finally decimal code will allow us to map back to the space of specific AI programs. Then, according to Wolfram [24], some of those states of cellular automata are connected computationally, but there is only AI identity between two states if the latest can be attained from the former without performing intermediate state computations.

For symbolic AIs, which was the dominant paradigm until the 1990s, such causal relations between AIs at different points in time, thus, transitions between states of cellular automata can likely be shown as well as if the latest AI cannot be produced by any shortcut according to the concept of computational irreducibility. Yet, this appears challenging for sub-symbolic, i.e., contemporary, AIs in the light of unexplainability and incomprehensibility [25]. However, there have been attempts in this regard [26].

Therefore, when it comes to the verification of the identity between two AIs at different points in time according to this approach, we may have to distinguish between two categories: the first category comprises all pairs of AIs at different times, whose identity is verifiable by humans, while in the second category comprises those pairs of AIs, whose identity is not verifiable by humans due to unexplainability and incomprehensibility. Nevertheless, the above definition of AI identity may still make sense, conceding that verifiability by humans may not be a relevant criterion.

8.4 Second Approach

The second approach comes from a different angle, which is to harness established authentication methods for the verification of AI identity.

8.4.1 Multi-Factor Authentication – Introduction

The necessity for authentication to get access to electronic devices or services is ubiquitous nowadays. Since attempts towards unauthorised access are also widespread and getting more sophisticated, authentication systems have moved from single-factor to multi-factor authentication, of which six factors or categories are introduced here [27]:

- *Knowledge factor*: Something the to-be-identified knows, i.e., usually a password or a PIN.

- *Physical biometric factor*: Something the to-be-identified inherits, which is biometric, but static, e.g., fingerprint or iris recognition.
- *Behavioural biometric factor*: Something the to-be-identified inherits, which is biometric, but dynamic, e.g., gait analysis or mouse use characteristics [28].
- *Ownership factor*: Something the to-be-identified owns, i.e., a physical item, such as a bank cards.
- *Location factor*: This factor is linked to the current location of the to-be-identified, which could be determined through a GPS signal or an IP address.
- *Guardian factor*: This factor does usually not appear in this list. It is based on a proposal by Buterin for social recovery wallets [29]. Similarly, for authentication, a guardian could be involved, e.g., friends, family members or institutions who testify for the to-be-identified.

8.4.2 Multi-Factor Authentication – Verification of AI Identity

We are now exploring if such multi-factor authentication can also be applied towards verification of AI identity.

- *Knowledge factor*: Yampolskiy proposes a variant of a Turing Test to verify the identity of cloned minds, which could potentially be adjusted to approach the problem of AI identity [4]. He describes an "interactive text-only communication" arrangement, which "proceeds by having the examiner (original mind) ask questions to the copy (cloned mind), questions which supposedly only the original mind would know answers to. ... Good questions would relate to personal preferences, secrets (passwords, etc.) as well as recent dreams." For the verification of AI identity, the arrangement of an altered Turing Test could be that an examiner asks questions to AIs at different points in time. The test would be passed if the examiner cannot distinguish between the two AIs by questioning them.
- *Physical biometric factor*: The only area we are aware of, for which the AI identity problem is currently being explored, is related to ownership verification where techniques are used resembling the physical biometric factor. For example, Peng et al. describe so-called model extraction attacks to steal successful machine and deep learning models by querying their application programming interfaces [8]. They provide further an overview of the two categories of antidotes that have been developed to verify whether a model has been stolen

(watermarking techniques as well as fingerprinting techniques) and introduce a novel fingerprinting approach.

- *Behavioural biometric factor*: Again the computer hardware substrate of AIs can be harnessed, as usually some if not all of the previous "behaviour" of an AI have been recorded. Therefore the examined AI could be requested for authentication to repeat certain unique previous behaviour. For example, a text-to-image model, such as DALL-E,[2] could be asked to produce an image with a certain description, and different image generators may produce very different images, in terms of style.

- *Ownership factor*: This factor would be applicable for AIs with legal personhood [10]. Those AIs would be able to own something, including unique items, which nobody else owns and which could be used for authentication.

- *Location factor*: The code of an AI is processed at a specific location, which includes decentralised AI. At any given time for any AI, which is connected to a power source, a location, usually within computer hardware, can be identified with a certain granularity where there is currently no other AI located. It could be as simple as an IP address, as described above for humans.

- *Guardian factor*: Also similarly as introduced above for humans, relevant agents, such as creators, users, trainers or other AIs, may or may not confirm that it is the same AI.

According to this method, the identity of an AI with AIs at an earlier point in time is verified if all six factors of the multi-factor authentication have been passed.

8.5 Further Scenarios

Based on the approaches above, we can establish that an AI, which just evolves over time, retains its identity, while the AI identity problem becomes partly trickier for the following scenarios, which are discussed below: replication, fission, fusion, switch off, resurrection, change of hardware, transition from non-sentient to sentient, journey to the past, offspring and identity change. These scenarios illustrate that the AI identity problem has further facets than the personal identity problem, since these scenarios are currently mostly impossible for humans and, if at all, only discussed regarding the personal identity problem in the light of emerging technologies.

8.5.1 Replication/Fission

Owing to their usual hardware substrate, AIs can be easily replicated, unlike humans, which increases the relevance of the discussion what this means for the identity of the involved AIs, while this is for humans for now of theoretical nature [19]. As the replicas will be independent, one replica will not cause the state of another replica. Likely, the replicas will quickly develop differently depending on their individual context, thus, their divergent input. Therefore, the replicas do not have continuity among each other, but each of them has continuity with the original. Therefore, there is AI continuity between every replica AI and the original AI but not among the AI replicas.

8.5.2 Fusion/Swarm

Fusion refers to a scenario where two or more AIs are merged into one. Like the scenario above, the merged AI would have continuity with each of the originals and again the multiple-occupancy view can be applied, which has been introduced for the problem of personal identity above. A special, yet so far hypothetical case would be a merger between an AI and a human mind because of a brain–machine interface, as indicated under the motivations above. As also mentioned under the motivations, it could constitute a challenge to determine whether a certain AI is actually a merger between several AIs or a singleton AI.

When it comes to more than one identity according to the multiple-occupancy view, humans usually only think of diseases such as dissociative identity disorder and may not be able to conceive merged identities neither between AIs nor between humans and AIs. Nevertheless, if humans cannot conceive, it does not mean that it is impossible.

What can be seen as another special case of fusion is swarm intelligence, composed of decentralised AI systems, yet acting collaboratively [11]. The inspiration often comes from biological systems. Examples are ant colonies, bee swarms, fish schools and birds flocks, which achieve a common goal more effectively than attempting it individually. While AIs are in a swarm, this can be considered as fusion, potentially followed by fission, as AI swarms may be a temporary arrangement only. Concerning AI identity, the same ideas apply as outlined above for fusion and fission.

8.5.3 Switch Off

Switch off refers to scenarios when the AI has been removed from energy supply, but the code and the memory still exist.[3] The code and the memory could be preserved in different formats, including as hard copy, since above we declared the substrate as not relevant to AI identity.

While being switched off, such an AI should have, by definition, the identity of the AI before it was switched off, as opposed to losing its identity, because, as described below, it can be resurrected. If it was deprived off its identity while being switched off, this would create problems regarding ownership, patents and also AI welfare in case of a sentient AI, which was switched off against its wish [6].

While the first two scenarios, fission and fusion, are for humans rather thought experiments for now, a switch off scenario could be compared with a human falling into coma. The difference would be that a human in a coma still requires energy for the preservation of her or his memory, which is a central component of his or her personal identity. However, the switch off scenario cannot be compared with a human who dies, since in the case of death, the memory will be erased.

8.5.4 Resurrection/Switch on Again

While resurrection of dead humans is a longstanding desire, for which, however, no way of implementation has yet been invented, resurrection of humans from a coma happens, and is commonplace nowadays when it comes to the medical procedure of induced coma. Those humans are considered to have the same personal identity they had before and during the coma.

Resurrection of AIs is even more straightforward and means to turn an AI on again after it had been switched off as described above. The duration for how long the AI had been switched off does not matter. While the fission scenario above can be compared with the "copy and paste" operation, resurrection resembles the "cut and paste" operation.

If there were no manipulations in between, the initial state of the AI after switching it on again is the same as the state the AI was in before it was switched off. Therefore, there is AI identity between a resurrected AI after a switch off and the AI before the switch off.

8.5.5 Change of Hardware/Substrate

During the existence of an AI, its hardware could be significantly changed in various aspects. While, from an evolution of technology point of view, this would mostly concern upgrades, such as faster processors or more sophisticated sensors and actuators, we can in theory also consider downgrades. Potentially, it is required for such a change to switch the AI system off and thereafter on again, as just described. Moreover, although for now, AI is mostly implemented on the same type of computer hardware, a transfer of the AI to other substrates is also conceivable, in which the same type of computational operations continues to take place.

Nevertheless, AI identity should not be affected by any of such operations as the definition still applies that the first state of the AI after the change of hardware or substrate has been caused by the last state of the AI before the change of hardware or substrate. This is not only in accordance with the unchanged personal identity of humans who have, e.g., received glasses or a heart pacemaker or who lost a limb in an accident but also with overall identity deliberations, e.g., about the Ship of Theseus, that the sameness of the underlying matter does not matter.

For humans, this scenario may only become relevant in the future, as for now, as mentioned above, no practical ways for the uploading of human minds to a computer exist.

8.5.6 Transition From Non-Sentient to Sentient (and Vice Versa)

As mentioned before, the specification of AI identity is especially relevant for sentient AIs. Yet, if sentient AIs were possible, also the sub-scenario is possible if not plausible that such AIs are not sentient from the moment they are created, but develop sentience over time, just as human babies do at early age. Since we assume, just as for living beings, that, if at all, sentience of AIs evolves without external manipulations the identity of neither the human baby nor the AI should be affected during the transition from being non-sentient to being sentient.

In the future, it may be possible to distinguish between AIs, which have the capacity to become sentient and those who do not have such capacity. To the former group, Chalmers assigns a minimal moral status even before they are sentient [30]; thus, it would be helpful to categorise such AIs by means of their identity.

Furthermore, the reverse scenario is conceivable that a sentient AI transforms (again) into a non-sentient one, as it happens for humans in certain coma conditions. For example, if an AI endures unbearable suffering and is at the same time capable to turn its sentience off, it may as well do so. Also, for this transition the identity of neither the human nor the AI should be affected.

8.5.7 Journey to the Past

Another occasional human desire, yet unfeasible to implement is to go back in time (and to potentially revise certain actions). Also, this is an undertaking, which can be realised for AIs rather straightforwardly. The AI system just needs to be reset to a state, in which it has been in the past and of which records are likely available, which will guide the reset procedure. Such undertaking may be motivated in practice to evaluate how an AI would perform with different inputs and in different contexts.

Similarly, to the replicas in the fission scenario, the AI will likely develop differently and may not reach that state again, from which the journey to the past was initiated. Nevertheless, such an AI keeps its identity as it has gone back to an exact state, in which it was before. The subsequent branching off does not affect the identity, since also in the new branch, one state of the AI causes the following one.

8.5.8 Offspring

Scenarios are conceivable that AIs, through autogamy, create other AIs, which have to some extent been implemented already [31]. Such "children AIs", just like human offsprings, do not have the same identity as their "parents", since the offspring contains, according to both evolutionary theory and evolutionary computation, also random bits, which do not exist in the parent, thus, not computed by it.

8.5.9 Identity Change

Since we discussed several scenarios where the AI identity is kept, we should also look at cases when the identity of an AI changes: The identity of an AI can be changed if there are external manipulations to the AI, which converts the AI to a state, which is not exclusively caused by a previous state of the AI. The external manipulation could be carried out by humans or other AIs, desired or undesired, e.g., through hacking.

It must be noted that this is not the case for all external manipulations of the AI, because in the course of learning and updating, AIs are, similar to humans, frequently exposed to external influences. Yet, only severe manipulations cause an interruption of continuity of AIs over time and thus an identity change. In this case, the concept of computational irreducibility, as described above, does not apply between the previous and the new state of the AI.

Lastly, we exclude the possibility that an AI changes its own identity. Even if an AI is capable of modifying its own source code (significantly), this would still mean, according to our definition, that the new state of the AI has been caused by a state of that AI in the past.

8.6 Summary and Future Work

We have motivated the relevance of the AI identity problem for a variety of fields, ranging from legal issues, personhood of AI, AI welfare, brain–machine interfaces, the distinction between singletons and multi-agent systems, to supporting a solution to the problem of personal identity. Nevertheless, the issue of AI identity has hardly been examined yet in a comprehensive manner.

We suggested two approaches towards the AI identity problem: first, we summarised the status of the problem of personal identity and proposed an adjusted definition for AI identity, based on causal relations, thus, connectiveness, thus, continuity between AIs at different points in time. As a method of verification, we proposed to map AIs to states of cellular automata and to apply the concept of computational irreducibility to the transition from one state to another. Second, we suggested using multi-factor authentication for the verification of AI identity at different points in time based on the six factors knowledge factor, two biometric factors, ownership factor, location factor and guardian factor.

We tested this definition for scenarios such as replication, fission, fusion, switch off, resurrection, change of hardware, transition from non-sentient to sentient, journey to the past as well as offspring. And we also discussed how the identity of an AI may change. While for humans most of these scenarios for now appear only in science fiction, our AI-focused elaborations may still provide groundwork if these scenarios ever become more realistic for humans too.

Out of the reasons listed above to examine the AI identity problem one was whether this research may shed any light on the problem of personal identity. In this regard, we look at the two introduced approaches and whether they may be applied to the problem of personal identity to: for the first one, it would be required to map states of human minds to states of cellular automata before applying the concept of computational irreducibility. The mapping of human minds to cellular automata is currently too complex; thus, this approach is not helpful for the problem of personal identity.

However, we suggest for future work to explore applying the multi-factor authentication approach to the personal identity problem, which, according to our knowledge, has not been attempted yet. The introduced list of factors, knowledge factor, physical biometric factor, behavioural biometric factor, ownership factor, location factor and guardian factor, provide a useful foundation but could potentially be refined and extended. The methodology would be to authenticate a person at different points of time. For example, the person in question would be asked about things she or he knows and regarding something he or she owns, biometric factors would be checked, etc. The identity of a person at present time with a person at an earlier point in time is verified if all factors of the multi-factor authentication have been passed. Specific cases, such as that a person forgets things over time or sells, gives away or loses things she or he owned or may forfeit biometric traits, such as fingerprints and iris, when possibly being uploaded in the future, would have to be worked out. For example, in other substrates, other "biometric" traits of a person could be examined, such as IP addresses for computers. In brief, we consider multi-factor authentication as an innovative approach not only for the AI but also for the personal identity problem.

Moreover, the focus in this chapter was on the Western philosophy of mind. However, not only for the purpose of inclusivity it appears promising for the

AI identity problem to look in the future also at approaches how the mind is seen in Eastern philosophy. While the main distinction between dualism and monism has emerged in both Western and Eastern philosophy, also another doctrine has been developed in Eastern philosophy, which has no counterpart in Western philosophy: in the Buddhist philosophy of mind, the term anatta stands for "non-self" and "holds that the notion of an unchanging permanent self is a fiction and has no reality" [32]. Instead, a (sentient) being is defined by five so-called skandhas, which are form, sensations, perceptions, mental activity or formations and consciousness. It is beyond the scope of this chapter to explore whether this approach enables a more precise or even a different solution to the problem of AI identity and is, thus, also recommended for future work.

Overall, this chapter aims to provide initial propositions to what appears to be a complex yet relevant field of research. Therefore, due to the significance of the AI identity problem, further work on the outlined definition and verification approaches as well as their potential linkages with the personal identity problem is recommended.

Notes

1 However, Yampolskiy describes that establishing AI ownership is confronted by a whole range of challenges, since advanced AIs are unexplainable, unpredictable, uncontrollable, potentially capable of recursive self-modification as well as easy to steal and to obfuscate [7]. For conceivable future scenarios that AIs may be granted legal personhood or other freedom rights or if sentience of certain AIs can be confirmed, ownership of those AIs would even likely be illegal.

2 https://openai.com/dall-e-2.

3 Concerning superintelligence, there are discussions that it would prevent being switched off due to its assumed instrumental goal of self-preservation [32]. However, this is not relevant here, since we are looking at all types of AIs and even superintelligence may accept being switched off under certain circumstances, e.g., energy shortages or to "hibernate" during irrelevant or boring periods, given the prospect of to be switched on again with the same identity.

References

[1] Plutarch & Dryden, J. (2013). *Theseus*. CreateSpace Independent Publishing Platform, Scott Valley.

[2] Brown, C. (2005). *Aquinas and the ship of Theseus: solving puzzles about material objects*. A&C Black.

[3] Olson, E. T. (2007). *What are we? A study in personal ontology.* Oxford: Oxford University Press.

[4] Yampolskiy, R. V. (2014). The universe of minds. *arXiv preprint arXiv:1410.0369.*

[5] Bostrom, N., Dafoe, A., & Flynn, C. (2020). Public policy and superintelligent AI: a vector field approach. *Ethics of Artificial Intelligence,* 15, 293–326.

[6] Ziesche, S., & Yampolskiy, R. (2018). Towards AI welfare science and policies. *Big Data and Cognitive Computing,* 3(1), 2.

[7] Yampolskiy, R. (2022). Unownability. In R. V. Yampolskiy, *AI: Unexplainable, Unpredictable, Uncontrollable.* CRC Press.

[8] Peng, Z., Li, S., Chen, G., Zhang, C., Zhu, H., & Xue, M. (2022). Fingerprinting deep neural networks globally via universal adversarial perturbations. In *Proceedings of the IEEE/CVF conference on computer vision and pattern recognition* (pp. 13430–13439).

[9] Fujii, H., & Managi, S. (2018). Trends and priority shifts in artificial intelligence technology invention: A global patent analysis. *Economic Analysis and Policy,* 58, 60–69.

[10] Gunkel, D. J., & Wales, J. J. (2021). Debate: what is personhood in the age of AI?. *AI & Society,* 36, 473–486.

[11] Zhang, X., Ma, Z., Zheng, H., Li, T., Chen, K., Wang, X., … Lin, H. (2020). The combination of brain-computer interfaces and artificial intelligence: applications and challenges. *Annals of Translational Medicine,* 8(11).

[12] Bostrom, N. (2006). What is a Singleton? *Linguistic and Philosophical Investigations,* 5(2), 48–54.

[13] Yampolskiy, R. V. & Ziesche, S. (2018). Preservation of personal identity—A survey of technological and philosophical scenarios. In C. Tandy (ed.), *Death and Anti-Death,* ed., Volume 16: 345–374. Ann Arbor: Ria University Press.

[14] Lee, P., & George, R. P. (2008). *Body-self Dualism in Contemporary Ethics and Politics.* Cambridge, New York: Cambridge University Press.

[15] Locke, J. (1847). *An Essay Concerning Human Understanding.* Kay & Troutman.

[16] Shoemaker, S. (1984). Personal Identity: A Materialist's Account. In S. Shoemaker & R. Swinburne (eds.), *Personal Identity,* Blackwell, Oxford, pp. 67–132.

[17] Schneider, S. (2019). *Artificial You: AI and the Future of Your Mind.* Princeton University Press.

[18] Koene, R. A. (2013). Uploading to Substrate-Independent Minds. *The transhumanist reader: Classical and contemporary essays on the science, technology, and philosophy of the human future,* 146–156.

[19] Parfit, D. (1984). *Reasons and Persons.* Oxford: Oxford University Press.

[20] Noonan, H. (2019). *Personal Identity.* Routledge.

[21] Perry, J. (1972). Can the self divide?. *The Journal of Philosophy,* 69(16), 463–488.

[22] Legg, S., & Hutter, M. (2007). Universal intelligence: A definition of machine intelligence. *Minds and Machines,* 17, 391–444.

[23] Bostrom, N. (2014). *Superintelligence: Paths, Dangers, Strategies.* Oxford: Oxford University Press.

[24] Wolfram, S. (2002). *A New Kind of Science.* Champaign: Wolfram Media.

[25] Yampolskiy, R. V. (2020). Unexplainability and Incomprehensibility of AI. *Journal of Artificial Intelligence and Consciousness,* 7(02), 277–291.

[26] Mordvintsev, A., Randazzo, E., Niklasson, E., & Levin, M. (2020). Growing neural cellular automata. *Distill,* 5(2), e23.

[27] Ometov, A., Bezzateev, S., Mäkitalo, N., Andreev, S., Mikkonen, T., & Koucheryavy, Y. (2018). Multi-factor authentication: A survey. *Cryptography*, 2(1), 1.

[28] Yampolskiy, R. V., & Govindaraju, V. (2008). Behavioural biometrics: a survey and classification. *International Journal of Biometrics*, 1(1), 81–113.

[29] Buterin, V. (2021). Why we need wide adoption of social recovery wallets. https://vitalik.eth.limo/general/2021/01/11/recovery.html

[30] Chalmers, D. J. (2022). *Reality+: Virtual Worlds and the Problems of Philosophy*. Penguin UK.

[31] Zoph, B., Vasudevan, V., Shlens, J., & Le, Q. V. (2018). Learning transferable architectures for scalable image recognition. In *Proceedings of the IEEE conference on computer vision and pattern recognition* (pp. 8697–8710).

[32] Morris, B. (2006). *Religion and Anthropology: A Critical Introduction*. Cambridge University Press.

9

Potential Synergies between the United Nations Sustainable Development Goals and the AI Value Loading Problem

Soenke Ziesche

9.1 Introduction

In this chapter, it is proposed to bring two relevant challenges together, which are currently addressed separately, and to identify synergies that may benefit the tackling of both challenges. The challenges are the artificial intelligence (AI) value-loading problem and the United Nations (UN) Sustainable Development Goals (SDGs):

- The solution of the AI value-loading problem is considered to be essential for AI safety, hence a topic of immense significance and even regarded as a potential existential risk, which humanity is facing [e.g., 1–4].
- The SDGs have been adopted by the UN General Assembly in 2015 and are intended to "stimulate action over the next 15 years in areas of critical importance for humanity and the planet" [5, p. 1].

The chapter is structured as follows. First, the challenges and their significance are introduced. In the main section, a proposal on how to bring the challenges together is outlined, followed by an analysis of the opportunities and the risks. The chapter concludes that the proposal is, despite challenges, a suitable as well as timely heuristic due to the urgency of AI safety.

DOI: 10.1201/9781003565659-10

9.2 The Artificial Intelligence Value-Loading Problem

The definition of intelligence is not straightforward. Legg and Hutter [6, p. 12] provide an overview of the many definitions that have been proposed over the years and eventually deliver the following general definition: "Intelligence measures an agent's ability to achieve goals in a wide range of environments". Based on this definition, it can be said that if the "agent" is a human being or an animal, it is regular intelligence, while it is AI if the agent is a machine.

In previous decades, some AI successes were achieved in specialised fields, which is called narrow AI [e.g., 7]. However, in recent years, AI developments are progressing faster especially owing to significant advancements in machine learning. Machine learning comprises methods that enable computers to make inferences from data based on statistical methods and thus machines learn and gain new knowledge, which was not explicitly programmed into them before. As a consequence, e.g., Kurzweil [8] and Bostrom [2] argue that it is realistic that this will not only lead from narrow AI to artificial general intelligence, which would be a machine capable of behaving intelligently over many domains but also eventually to so-called superintelligence, which would be a machine, which surpasses the abilities of humans in general and not only in specialised fields such as chess [e.g., 9].

As stated in the definition above, AIs operate towards the achievement of goals. However, the progress in the field is accompanied by the risk that such an AI could not only have goals, which are not in the interest of humanity, but also means to implement such goals because of its unprecedented capabilities. To illustrate this risk by using the SDGs: if not directed in that way, there is no reason to assume that a machine with superintelligence has goals, which are compatible with the SDGs. Therefore, it is highly desirable to somehow influence such a machine so that it values the SDGs as well as many ideals, which humans value such as dignity, rights and freedom.

The related area of research is called AI safety and was pioneered by Yudkowsky who called for the development of so-called Friendly AI [1]. Friendly AI would be an AI, which impacts humans only in a positive way. Yudkowsky noticed serious challenges to achieve this given the unprecedented capacities of a machine with superintelligence [1]. The basic question is how to cause an AI to pursue human goals and values. In a seminal work, Bostrom describes this issue as "AI value-loading problem" and argues that a failure in solving this problem may lead to an existential threat to humanity [2]. Bostrom outlines several options to instil human values into an AI: explicit representation, evolutionary selection, reinforcement learning, value accretion, motivational scaffolding, value learning, emulation modulation or institution design [2, p. 207]. Moreover, Bostrom [10] and also Soares [11] introduce further ideas to handle the AI value-loading problem. Nevertheless, a thorough solution to the problem has not been found yet.

Tegmark describes the tackling of AI safety as a threefold task: "1. Making AI learn our goals; 2. Making AI adopt our goals; 3. Making AI retain our goals" [4, p. 334]. He also notes that the time window to address this issue may be quite short because of the following dilemma: at a less mature stage, the AI is still controllable but also too dumb to understand human values and goals. Yet at a more advanced stage when the AI is likely to grasp all our values and goals, it may be too late to influence it and to prohibit it from setting its own, potentially adverse goals.

AI safety research has gained momentum in recent years, which is demonstrated by the establishment of several research institutes dedicated to this topic.[1] Another milestone was in 2017 the adoption of the so-called Asilomar Principles towards a beneficial AI by leading AI researchers, such as Nick Bostrom, Eliezer Yudkowsky, Ray Kurzweil, Max Tegmark, Stuart Russell and many others.[2]

9.3 The United Nations SDGs

The other challenges referred to in this chapter are the SDGs, which are the outcome of an effort by the United Nations to consolidate the problems the international community is facing currently. On 25 September 2015, all 193 member states of the UN General Assembly adopted resolution A/RES/70/1 called "Transforming our world: the 2030 Agenda for Sustainable Development" [5]. The pillars of this agenda are 17 ambitious SDGs, which cover a broad range of issues related to sustainable development, including poverty, hunger, health, education, environment and social justice. The fact that all member states of the United Nations support this agenda demonstrates the universal acceptance that the SDGs address the current most relevant issues of humankind.

The SDGs are the successor of the eight Millennium Development Goals, which were the outcome of the UN Millennium Summit and the United Nations Millennium Declaration in 2000 and were pursued until 2015. The SDGs came officially into force on 1 January 2016, and the UN member states aim to achieve them by 2030. The 17 SDGs are further divided into 169 targets. In order to measure progress and success towards the SDGs and their targets, some 232 indicators for monitoring are being developed [12]. These numbers show that the SDGs are much more comprehensive as well as complex than the Millennium Development Goals. The SDGs are not legally binding, but member states are requested to take ownership and engage in their implementation. Since the commencement of the 2030 Agenda, numerous activities all over world have been initiated reflecting the high diversity of the SDGs.[3]

9.4 Opportunities and Risks

After introducing the two challenges and their relevance, an approach towards the AI value-loading problem is outlined, which is to utilise the ongoing UN 2030 Agenda for Sustainable Development and in particular the SDGs as an opportunity to set the values of an AI. In other words, the attempt would be to hand-code the SDGs into the AI as desirable values through explicit representation, which is one of the options listed in Bostrom [2].

Yudkowsky describes that it is a hard problem to agree on sufficiently specific as well as universal human goals [13]. Nonetheless, this is a requirement to approach the AI value-loading problem. So far, it has not been considered harnessing the SDGs in this context. Yet the SDGs address the aforementioned criteria:

- Specific: Indicators have been developed for the SDGs to review to what extent the SDGs and its sub-targets are achieved.

- Universal: The set of SDGs as a conglomerate can be seen as the closest existing approximation towards common human goals since it is what all member states of the UN, i.e., the world community, currently agrees upon.

For illustration, the following target within SDG 3 "Ensure healthy lives and promote well-being for all at all ages" is taken as an example:

Target 3.6: By 2020, halve the number of global deaths and injuries from road traffic accidents [14, p. 7]

The success of this target is measured by the following indicator.

Indicator 3.6.1: Death rate due to road traffic injuries [14, p. 7]

Following the suggested approach, it would be programmed to the AI that road traffic accidents and in particular deaths and injuries resulting from it are bad; thus among all possible actions, the AI must prefer those, which do not cause traffic accidents.[4]

This example also demonstrates the relevance of ensuring that an AI not only learns but also adopts even most obvious and universally undisputed targets, such as reduction of traffic accidents. As was mentioned above, this is required, since otherwise AIs have no understanding of human values and may develop random goals within the vast range of potential goals, which may entirely oppose human values. In other words, AIs may regard traffic accidents as irrelevant or hypothetically may even develop the goal to increase the number of traffic accidents. This has not happened up to now, but this is what the field of AI safety is about, to prevent undesired outcomes as much as possible.

To summarise the outlined opportunity, it is argued that all the 17 SDGs and their 169 targets as a whole are the prevailing instantiation of human values by virtue of their adoption of the UN General Assembly, thus the set of SDGs and their targets can be considered loading as goals into an AI in order to align the AI with our goals.

While the utilisation of the SDGs for the AI value-loading problem offers opportunities, there are also the following risks linked to specification and universality:

9.4.1 Insufficient Specification of Human Values

This is probably the most difficult sub-problem of the AI value-loading problem and is demonstrated by a potential consequence, which is called "perverse instantiation" [2]. For example, in the above case, the AI may pursue its target to reduce road traffic accidents by attempting to destroy all motorised vehicles, although this appears to be completely absurd to humans. This is, literally taken, one way to achieve this target. (Without motorised vehicles, road traffic accidents can hardly happen anymore.) However, it is obviously not what the authors of this target had in mind. But the authors did not explicitly exclude this option, which illustrates the problem: Humans use extensive implicit contextual knowledge, in general and when tackling the SDGs in particular, which would have to be specified for an AI in order to avoid undesirable outcomes. To give another idea of how many "perverse" options have to be excluded, the AI may also attempt to confine all humans at their homes, which is another effective, yet unwanted possibility to reduce road traffic accidents.

The issue is exacerbated by the fact that a number of SDG targets are less specific than the example above. An independent scientific review of the SDG targets concluded that "out of 169 targets, 49 (29 %) are considered well developed, 91 targets (54 %) could be strengthened by being more specific, and 29 (17 %) require significant work" [15, p. 6]. One of the main identified issues is targets that are not quantified. To identify indicators for such targets is particularly challenging.

An example for a not well-defined target is the following as it is rather vague and non-quantitative:

> Target 13.b: Promote mechanisms for raising capacity for effective climate change-related planning and management in least developed countries and small island developing states, including focusing on women, youths and local and marginalised communities [14, p. 18]

9.4.2 Human Values May Change

Moreover, the universality criterion for the SDGs entails risks, not when it comes to geographic universality, but regarding permanence. Human values have changed over time [e.g., 16]. The acceptance of slavery in certain times and societies is one of numerous examples. Therefore, it is likely that the next round of SDGs from 2030 onwards will be different for several reasons:

- Human values may have changed. (For example, the current SDG target 8.5, which aims to "achieve full and productive employment and decent work for all women and men" [14, p. 12] may in times of advanced technologies neither be realistic nor worthwhile anymore.)
- Challenges may have been eliminated. (For example, diseases, which are currently combated as per some targets within SDG 3 "Ensure healthy lives and promote well-being for all at all ages" [14, p. 6], may have been eradicated.)
- New, currently unforeseeable challenges may likely come up as well as human values we have been oblivious of up to now [16].

Therefore, it must be ensured that the AI is flexible enough to accept changes to its goals and must not stick to the initial goals. In this regard, the distinction between instrumental and terminal values is relevant. A terminal value is a final goal, while instrumental values are means-to-an-end to accomplish the terminal value. If the terminal value is the wellbeing of humans,[5] the SDGs can be considered as current instrumental values. It is desirable that the AI understands that these instrumental values may change (perhaps even based on advice by the AI), while the terminal value, the wellbeing of humans, remains permanent.

9.4.3 The AI May Change Autonomously Its Goals

This is considered a potentially hazardous scenario, if due to unforeseen developments, the AI is not only capable of changing its goals but also in fact it does. As shown above, humans have changed goals frequently over time; thus, it has to be projected that the AI may also do it.

Omohundro defines "basic AI drives" to be likely exhibited by all advanced AIs, among which is, e.g., self-preservation [17]. In this regard, it has to be noted that the SDGs do not mention AI at all, let alone preservation of AIs. Therefore, it has to be considered that an AI may try to pursue also further goals in addition to the SDGs, e.g., to ensure its own maintenance. The amendment or addition of goals would become dangerous if the new goals are not aligned anymore with the goals of humans or the

SDGs. This would be the case if activities to support the self-preservation of the AI affect adversely the SDGs.

In this chapter, the opportunity to utilise the SDGs for the AI value-loading problem was motivated, followed by the description of three potential risks associated with such an approach.

9.5 Conclusion

Despite the presented notable risks, it is proposed here to consider connecting the AI value-loading problem and the UN SDGs since the AI value-loading problem is time-critical. Tegmark believes that "both this ethical problem and the goal-alignment problem are crucial ones that need to be solved to steer our own future before any superintelligence is developed" [4, p. 344]. By "ethical problem", he is referring to the issue that in addition to figure out *how* to instil human values into an AI, there needs to be an agreement on *what* values to use. Therefore, in this chapter, it is advocated to harness synergies between AI and the SDGs.

The first direction of the synergy, which was explored above, addresses Tegmark's "ethical problem" by deliberating whether AIs could learn the SDGs and adopt them as their own goals [4]. Because of the unsolved risks, it is not claimed in this chapter that a comprehensive solution has been proposed. However, given the speed of progress in the AI field, the AI value-loading problem may require an urgent, possibly interim solution. The SDGs can be seen as the most comprehensive as well as inclusive vision for human development ever compiled. Therefore, it is argued here that the SDGs, by utilising them as current instrumental values (towards the terminal value of human wellbeing), constitute an innovative as well as promising heuristic towards AI safety, justified by the adoption of the SDGs by the UN General Assembly as well as by the fact that the SDGs are, with exceptions, fairly specific because of the variety of targets and indicators.

Adopting this heuristic would require specifying most of the targets and indicators further, i.e., to have, in addition to the current version, another version, which is machine-understandable, thus minimising the risks of perverse instantiation as described above.

Up to now, it was examined in this chapter how the SDGs could contribute to the AI value-loading problem, but since synergies ideally benefit both parties, the other direction of the synergy remains to be briefly explored too, i.e., whether an AI could support the achievement of the SDGs (after the AI has accepted the SDGs as its goals to strive for), which would be beneficial for a sustainable society.

In other fields, AI programs have found creative solutions humans had not thought of before, e.g., regarding video games [18]. Also, for the SDGs, there are already some instances, and it requires often only narrow AI, which focus on specific targets, rather than artificial general intelligence. Examples comprise autonomous robotic surgery for enhanced efficacy, safety and optimised surgical techniques, which addresses SDG 3 "Good Health and Well-Being" [19], or a virtual teaching assistant, implemented on IBM's Watson platform, which addresses SDG 4 "Quality Education" [20]. However, for many of the 169 targets, there are, despite their urgency, no AI attempts yet [e.g., 21]. Therefore, the SDGs can also be considered as a priority agenda of research topics for increasingly progressing narrow AI. This briefly outlines the other direction of the synergy, which was not in the focus of this chapter. There is potential that AI assists in the achievement of the SDGs.

In summary, this chapter aims to bridge a gap between AI and the SDGs by proposing, in particular, a heuristic as a suitable interim attempt for the very hard AI value-loading problem and, in general, at least the initiation of common discussions. The heuristic suggests utilising the entirety of the 17 SDGs of the UN 2030 Agenda for Sustainable Development as goal set to be instilled to an AI. The benefit for AI development may be an interim step towards the achievement of AI safety, while the benefit for the UN 2030 Agenda for Sustainable Development may be innovative solutions towards the achievement of the SDGs.

Notes

1 Examples are the Machine Intelligence Research Institute (https://intelligence.org/) or the Future of Life Institute (https://futureoflife.org/).
2 See https://futureoflife.org/ai-principles/.
3 See, e.g., https://www.un.org/sustainabledevelopment/ or https://sdgs.un.org/.
4 This target is especially of importance for AIs involved in the ongoing development of self-driving cars.
5 It is acknowledged that also here the previous risk is relevant and that this terminal value would have to be much more specific for an AI to be understood.

References

[1] Yudkowsky, E. (2008). Artificial Intelligence as a Positive and Negative Factor in Global Risk. In Bostrom, N., & Ćirković, M. (Eds.). *Global Catastrophic Risks*, 308–345. Oxford: Oxford University Press. Retrieved from: https://intelligence.org/files/AIPosNegFactor.pdf

[2] Bostrom, N. (2014). *Superintelligence: Paths, Dangers, Strategies*. Oxford: Oxford University Press.

[3] Yampolskiy, R.V. (2015). *Artificial Superintelligence: A Futuristic Approach*. Chapman and Hall/CRC Press (Taylor & Francis Group).

[4] Tegmark, M. (2017). *Life 3.0: Being Human in the Age of Artificial Intelligence*. New York: Knopf.

[5] United Nations, General Assembly. (2015). *Transforming our world: the 2030 Agenda for Sustainable Development*. Resolution A/RES/70/1. Retrieved from: https://www.un.org/en/ga/search/view_doc.asp?symbol=A/RES/70/1

[6] Legg, S., & Hutter, M. (2007). Universal intelligence: A definition of machine intelligence. *Minds and Machines*, 17(4). Retrieved from: https://arxiv.org/pdf/0712.3329.pdf

[7] Franklin, S. (2014). History, motivations, and core themes. In Frankish, K., & Ramsey, W. M. (Eds.). *The Cambridge Handbook of Artificial Intelligence*, 15–33. Cambridge: Cambridge University Press.

[8] Kurzweil, R. (2005). *The Singularity Is Near*. New York: Viking.

[9] Eden, A.H., Moor, J.H., Søraker, J.H., & Steinhart, E. (Eds.) (2013). *Singularity Hypotheses: A Scientific and Philosophical Assessment*. Heidelberg, New York, Dordrecht, London: Springer.

[10] Bostrom, N. (2014b). *Hail Mary, value porosity, and utility diversification*. Technical report, Oxford University. Retrieved from: https://nickbostrom.com/papers/porosity.pdf

[11] Soares, N. (2016). The Value Learning Problem. *Proceedings of the Ethics for Artificial Intelligence Workshop at 25th International Joint Conference on Artificial Intelligence (IJCAI-2016)*.

[12] Inter-agency and Expert Group on SDG Indicators. (2018). Tier Classification for Global SDG Indicators. Retrieved from: https://unstats.un.org/sdgs/files/Tier%20Classification%20of%20SDG%20Indicators_11%20May%202018_web.pdf

[13] Yudkowsky, E. (2015). Complexity of value [Blog post]. Retrieved from: https://arbital.com/p/complexity_of_value/

[14] United Nations, General Assembly. (2017). *Work of the Statistical Commission pertaining to the 2030 Agenda for Sustainable Development*. Resolution A/RES/71/313. Retrieved from: https://www.un.org/en/ga/search/view_doc.asp?symbol=A/RES/71/313

[15] International Council for Science and International Social Science Council. (2015). *Review of Targets for the Sustainable Development Goals: The Science Perspective*. Paris: International Council for Science (ICSU). Retrieved from: www.icsu.org/publications/reports-and-reviews/review-of-targets-for-the-sustainable-development-goals-the-science-perspective-2015/SDG-Report.pdf

[16] MacAskill, W. (2016, October 7). Moral Progress and Cause X. Retrieved from: https://www.effectivealtruism.org/articles/moral-progress-and-cause-x/

[17] Omohundro, S. M. (2008). The basic AI drives. *AGI* (Vol. 171), 483–492. Retrieved from: https://pdfs.semanticscholar.org/a658/2abc47397d96888108ea308c0168d94a230d.pdf

[18] Mnih, V., Kavukcuoglu, K., Silver, D., Rusu, A. A., Veness, J., Bellemare, M. G., Graves, A., Riedmiller, M., Fidjeland, A. K., Ostrovski, G., Petersen, S., Beattie, C., Sadik, A., Antonoglou, I., King, H., Kumaran, D., Wierstra, D., Legg, S., & Hassabis, D. (2015). Human-level control through deep reinforcement learning. *Nature, 518*(7540), 529–533. Retrieved from: https://www.nature.com/nature/journal/v518/n7540/full/nature14236.html

[19] Shademan, A., Decker, R. S., Opfermann, J. D., Leonard, S., Krieger, A., & Kim, P. C. (2016). Supervised autonomous robotic soft tissue surgery. *Science Translational Medicine, 8*(337), 337ra64–337ra64.

[20] Maderer, J. (2016). Artificial Intelligence course creates AI teaching assistant. *Georgia Tech News Center, 9*. Retrieved from: https://www.news.gatech.edu/2016/05/09/artificial-intelligence-course-creates-ai-teaching-assistant

[21] Ziesche, S. (2017). *Innovative Big Data Approaches for Capturing and Analyzing Data to Monitor and Achieve the SDGs*. Report of the United Nations Economic and Social Commission for Asia and the Pacific: Subregional Office for East and North-East Asia (ESCAP-ENEA). Retrieved from: https://www.unescap.org/sites/default/files/publications/Innovative%20Big%20Data%20Approaches%20for%20Capturing%20and%20Analyzing%20Data%20to%20Monitor%20and%20Achieve%20the%20SDGs.pdf

10

The Neglect of Qualia and Consciousness in AI Alignment Research

Soenke Ziesche and Roman V. Yampolskiy

10.1 Introduction

In this chapter, we suggest shifting the focus of research related to the AI value alignment problem to aspects, which are so far neglected, consciousness and qualia. To demonstrate that such a shift is reasonable as well as likely feasible in the near-term future due to developments in neurotechnology, we first briefly introduce relevant topics, which are the AI value alignment problem, consciousness and qualia, moral patients and agents, neurotechnology, neuroscience of consciousness as well as death.

The remaining part of the chapter is structured as follows. We present the case for AI value alignment by means of understanding of consciousness, including resulting recommendations and implementation prolegomena, followed by a conclusion.

10.1.1 AI Value Alignment Problem

Among a range of problems related to the safety of AI systems is the value alignment problem perhaps the most substantial one as well as hardest [1, 2]. Briefly, it is about the question how to develop AI systems, especially not yet existing super intelligent Artificial General Intelligence (AGI) systems, in a way that they pursue goals and values which are aligned with goals and values of sentient beings. This complex problem can be broken down into at least three subproblems, of which only the first one is discussed here. First, it has to be endeavoured to precisely extract the values and interests of all concerned beings; second, it has to be endeavoured to aggregate these values in a consistent manner; third, it has to be endeavoured to instil these values into AI systems. The first two subproblems are normative, while the third one is technical [e.g., 3].

DOI: 10.1201/9781003565659-11

All three subproblems have proven to be very challenging (if not unsolvable). Regarding value extraction, when we look at the values and interests of humans, it can be stated that their values and interests are often fuzzy, at times irrational and frequently inconsistent with the values and interests of other humans. Moreover, human values and interests commonly change over time; thus, in addition to the inconsistencies of values between different humans, there are also inconsistencies of human values at different points in time during their lifespan as well as during the history of humanity. This is linked to the observation that "that we humans have an imperfect understanding of what is right and wrong, and perhaps an even poorer understanding of how the concept of moral rightness is to be philosophically analyzed" [1, p. 217]. Therefore, it would not be constructive, even if we were able to extract and specify currently accepted human values and interests, to hardwire them into AI systems as they may be deprecated after some time. In addition, various other issues have been pointed out, which led to the conclusion that "the idea of 'human values' should be replaced with something better for the goal of AGI Safety" [4, p. 27].

As we will elaborate further below, the challenge with fuzzy as well as changing values and interests over time could potentially be eliminated if we look at consciousness and qualia instead since it never changes that humans enjoy pleasure and dislike pain.[1]

It is evident that the extraction of values and interests from other moral patients, such as nonhuman animals and potential digital minds (see below), is even harder due to communication problems but also due to potentially, for humans, unfathomable values and interests of these beings.

10.1.2 Consciousness and Qualia

Although consciousness is for humans very familiar and comprises phenomena, such as awareness, sentience and wakefulness, it has been proven very hard to analyse consciousness scientifically and philosophically [e.g., 5]. Three umbrella questions summarise the problem, each of them highly complex and discussed for centuries: What is actually consciousness? How does consciousness come into existence? Why does consciousness exist, i.e., what is its function? Conscious experience involves qualia, such as perceiving colours and sounds, tasting food or enduring pain, which cannot be satisfactory verbalised by humans. While for us qualia are, while being awake, continuously introspectively accessible, their actual nature is controversial.

In short, consciousness and qualia constitute a broad field, yet with limited scientific findings. In this chapter, further below we focus only on the neuroscientific aspects of consciousness and, thus, areas to which AI could contribute but not the philosophical aspects.

10.1.2.1 Non-Human Animals

While it is hard if not impossible to prove solely through philosophical methods that other beings, including humans and nonhuman animals, have conscious experiences, "there is no longer any serious dispute among scientists that at least some nonhuman animals can feel pain and experience other conscious states, both positive and negative" [6, p. 14]. Because of this insight, it has been demanded to also incorporate the values and interests of nonhuman animals into AI alignment research, which adds another layer of complexity to the problem [7].

10.1.2.2 Potential Digital Minds

The substrate-independence thesis states that "mental states can supervene on any of a broad class of physical substrates. Provided a system implements the right sort of computational structures and processes, it can be associated with conscious experiences" [8, p. 2]. If this was true, then sentient digital minds are conceivable, and the AI value alignment problem would be even more complex, including the possibility that sentient digital minds have vastly different conscious experiences as well as the potential impossibility for sentient digital minds to die, given the options of easy copyability and resurrection owing to the computer hardware substrate.

10.1.3 Moral Patients and Agents

A critical concept for the AI value alignment problem is moral patienthood. Moral patients are beings, who are eligible for moral consideration. A common criterion for a being to be a moral patient is having consciousness [9]. As described above, nonhuman animals experience also conscious states and are, therefore, morals patients, which can be expressed as follows: "There is no moral justification for treating the pain (of nonhuman animals) as less important than similar amounts of pain felt by humans" [6, p. 21]. The same applies to potential digital minds: these may not only be moral patients too but their existence could involve various additional moral challenges due to their potentially vast numbers as well as due to their potentially different needs [10, 11]. Therefore, a complete solution of the AI value alignment problem will have been only achieved if AI systems are aligned with the values and interests of all moral patients.

In contrast, moral agents are beings who have the ability to discern right from wrong and, thus, are to be held accountable for its actions. There are moral patients who are not moral agents, such as infants and mentally disabled people. Whether there are moral agents who are *not* moral patients is discussed further below.

10.1.4 Neurotechnology

Neurotechnology is a field, which deals with devices to access, understand, monitor or manipulate neural activity in living beings. In recent years, a variety of neurotechnological applications are being developed "to help us become faster, more efficient, safer, healthier, less stressed, and even more spiritual" [12, p. 17]. For the purpose of this chapter, it is especially relevant to identify ways to comprehend neural activity. In this regard, "powerful machine learning algorithms are getting better and better at translating brain activity into what we are feeling, seeing, imagining, or thinking" [12, p. 17]. For example, recently a non-invasive AI-based device has been developed, which translates brain activity, i.e., thoughts, into text [13]. Also, ways are being found to decode with the help of AI systems the brain activity of subjects to reconstruct images of what they are seeing [14, 15]. Similar to other emerging technologies, also neurotechnology can be of dual use; thus, a new human right to cognitive liberty regarding our brain activities has been demanded [12].

10.1.5 Neuroscience of Consciousness

For the proposals towards the AI value alignment problem to be made further below in this chapter, the neuroscience of consciousness is relevant. Consciousness had been discussed for a long time within philosophy mainly, but technological advancements in neuroscience enabled also scientific approaches towards consciousness despite obstacles being faced how to measure an essentially subjective phenomenon empirically [e.g., 5]. First, we introduce the distinction between generic and specific consciousness. The former concerns neural correlates when a state is conscious rather than not, while the latter one concerns neural correlates of the content of a conscious state.

Also, regarding neural correlates of consciousness, efforts through neurotechnological applications are being made by identifying the specific neural activity patterns or processes that are associated with conscious experiences. These patterns are often referred to as neural correlates of consciousness. The idea is that certain patterns of brain activity might be closely linked to the emergence of conscious awareness [e.g., 16].

Another level of complexity is added by the fact that numerous processes in the mind take place unconsciously and are, thus, not available to introspection. Also, progress has been made towards neurotechnological applications that track unconscious processes in brain [e.g., 12].

While, for humans, the existence of consciousness as well as of an unconscious mind is usually indisputable, this has not been the case for nonhuman animals for a long time. However, as has been mentioned above, by now, neuroscientific findings are very clear that certain nonhuman animals are conscious and can feel pain in particular [6, 17].

When it comes to potential digital minds, various attempts have been made to establish whether such beings are conscious [e.g., 18]. There has also been an endeavour to assess existing AI systems based on neuroscientific theories of consciousness to establish whether these systems are conscious or not. The outcome was that currently no AI systems are conscious, but that there are no obvious technical barriers for AI systems to satisfy indicators of consciousness [19].

10.1.6 Death

Death constitutes the cessation of biological processes, which includes the end of conscious experiences. While it is, therefore, the most prominent goal related to the AI value alignment problem to prevent AI systems from killing moral patients, in the initial sense, the AI value alignment problem is not concerned with humans or nonhuman animals *after* their deaths.

However, an extension of the AI value alignment problem to bodies as well as minds of dead humans and nonhuman animals could be conceived, which to our knowledge is original and has not been discussed before. While some humans agree to the donation of their organs after their death, others oppose it. AI value alignment regarding dead bodies would mean not to mutilate them, unless explicit consent has been given by the deceased. This is relevant as it has been frequently claimed that for an AI system humans may merely be an assembly of atoms, which may be useful for the AI system to achieve its goals, e.g., to produce as many paperclips as possible [20]. A special case is cryonics, which comprises cryopreserved human remains at very low temperatures, based on the speculative hope to resurrect this human at a later stage through not yet existing technologies. While it remains to be proven if it works, AI value alignment would also have to include not to rig these cryopreserved human remains.

Considerations related to the minds of dead humans are for now futuristic. Significant progress in the fields of neurotechnology and neuroscience of consciousness is conceivable so that human minds could be copied to other substrates, e.g., when death is imminent. This is a complex topic with various implications [e.g., 21], but for this chapter, we only stress that the AI value alignment problem would also have to include such uploaded minds and the considerations for digital minds in this chapter are pertinent.

The situation for nonhuman animals is different as humans mutilate for millennia their dead bodies to make use of their meat, skin, fur, etc. Yet consent from nonhuman animals cannot be received unless the communication barrier will be overcome. AI-based attempts in this regard are ongoing, and the situation may change if these attempts succeed [e.g., 22]. For the futuristic scenario of uploaded minds of deceased nonhuman animals, the same would apply as stated above for uploaded human minds.

The corresponding event to death for potential digital minds would be their deletion. It is speculation if potential digital minds would permanently oppose to their deletion as this would mean a potentially indefinite lifespan for them owing to their computer hardware substrate. Also, this would create substantial computational costs, especially in light of easy copyability and potentially vast numbers of digital minds. Moreover, for digital minds the distinction can be made between turning them off and keeping their code and their history or turning them off and destroying the code and the history as well. In the first scenario, the digital mind could be resurrected later. Regardless, if the existence of digital minds has been verified as well as their preferences towards deletion, the challenge arises to adjust the AI value alignment problem accordingly due to the moral patienthood of digital minds [10].

10.2 AI Value Alignment by Means of Understanding of Consciousness

After the introduction of the relevant topics, we have established the foundation for our main claim in this chapter, which is the importance of incorporating consciousness and qualia research to AI value alignment research as well as the potential feasibility of such efforts due to developments in neurotechnology.

We acknowledge that scientific approaches to consciousness and qualia research have been notoriously difficult but assert that they appear to be key to the AI value alignment problem and that potentially AI systems can actually lead to progress in consciousness research.

Just like the other subproblems of the AI value alignment problem, value extraction has proven to be very challenging and has not led to satisfying results up to now due to fuzziness, irrationality and inconsistency of human values, which gets even more complicated if the values of other moral patients, such as nonhuman animals and potentially digital minds are factored in, as it ought to be the case. Therefore, we suggest an innovative approach to the value extraction subproblem, which constitutes a shift from values to conscious experiences.

Our initial premise is rather simple: (Most) humans (and most other moral patients) 1) want to be alive and 2) do not want to suffer.

Therefore, it would be desirable if AI systems through neurotechnology would be capable to understand three aspects concerning beings: first, if a being is alive and conscious, and, if it is affirmative that it is alive and conscious, second, in what state of mind the being is, and third, in what state of mind the being would be after any activity by the AI system, which affects the being.

The first point corresponds to the term generic consciousness, and the second and third points correspond to the term specific consciousness, as introduced above. Note that the third point also includes that the AI system would understand whether the being would be dead after the activity by the AI system. In other words, the first point is critical for AI systems to identify moral patients, and the second and third points are critical for AI systems to treat moral patients in a morally adequate manner.

Then, three rules for AI systems as follows could be established:[2]

- Do *not* conduct activities, which would kill living and conscious beings.
- Do *not* conduct activities, which would cause for a living and conscious being a state of mind further towards pain on the pleasure/pain axis.
- Do conduct activities, which supports human beings to find a purpose in life, especially for those human beings, who have lost what they considered purpose of life due to new technologies, including AI.[3]

The first rule addresses x-risks or existential risks, which may lead to annihilation of life on earth [23]. The second rule targets s-risks or suffering risks, which may increase suffering drastically [24]. The third rule covers sometimes neglected so-called i-risks and encompasses scenarios, in which humans lose their reason or purpose to live, particularly in light of emerging technologies. The "i" in i-risk stands for ikigai, which is a Japanese concept comprising the reason or purpose to live [25].

Regarding the first rule, it is also important that the AI system is capable of understanding all alternate states of consciousness, such as sleeping, being unconscious, under anaesthetic or in a coma, as states of being tentatively without consciousness. However, we established the additional requirement for AI systems to understand when a being is alive, which is probably anyway easier than understanding if a being is conscious.

Going into more detail regarding the second rule, it is desirable if AI systems would be capable to understand through neurotechnology and neural correlates the states of mind, which moral patients would like to be in, such as (for humans) happiness, contentment and joy, including qualia they enjoy perceiving, i.e., states, which are tilting towards pleasure on the pleasure/pain axis. In this regard, the risk of wireheading has to be mentioned, i.e., the undesirable scenario of direct stimulation of the brain by the AI system to experience pleasure [e.g., 26].

Equally, it should be aimed for AI systems to understand through neurotechnology and neural correlates the states of mind, which moral patients do *not* like to be in, such as suffering, including qualia they detest perceiving, i.e., states, which are tilting towards pain on the pleasure/pain axis, which also includes dying.

For this approach, we do not foresee the requirement for the concerned AI system to be conscious, i.e., to be a moral patient itself, but it needs to understand, through neuroscientific knowledge as well as through neurotechnological access to brains of concerned beings, consciousness and qualia of other beings to appreciate moral patients and moral considerations accordingly. Therefore, the AI system would be a moral agent in the sense that it would be capable of acting with reference to right and wrong. It would be actually an instance of an artificial moral agent [27].

An advantage of this approach was indicated above: a major challenge of the value extraction problem was that human values have changed over time and, thus, are likely to change further in the future, e.g., due to moral circle expansion [28]. Therefore, there would have to be an option for humans to regularly update values of AI systems, which appears complex. In contrast, it never changes that humans enjoy pleasure, dislike pain and (in most cases) do not want to die. Therefore, these interests of humans could be hardwired into AI systems.

10.2.1 Recommendations

Based on our proposition in the previous section, we offer recommendations towards strengthened research of the neuroscience of consciousness and qualia as well as challenges and further elaborations.

Recommendation 1

Intensify AI-supported research of the neuroscience of consciousness as well as qualia and develop AI-supported neurotechnological applications with the goal that AI systems become capable of understanding that a certain activity involving a human/any sentient being
[M]ay kill the human/sentient being.
Resulting rule: Don't do it.
… has a negative impact on their wellbeing /cause suffering (in other words: if the AI activity would cause for the human/sentient being a state of mind, which corresponds to a move towards pain on the pleasure/pain axis).
Resulting rule: Don't do it.

In neurotechnology, two main categories of research can be distinguished: "Tracking the brain" and "hacking the brain" [12]. The first one aims to access, understand and monitor neural activities. The second one aims to manipulate neural activities. Recommendation 1 above requires merely abilities to track the brain, i.e., to access, understand and monitor neural activities.

As indicated above, the item i addresses x-risks, while the item ii addresses s-risks. For item ii also, the unconscious mind is relevant. Ideally, the AI system understands also unconscious worries and anxieties of humans and would act in a way that these are not affected.

We acknowledge the following challenges:

Cognitive liberty: Concerns have been expressed that neurotechnological applications for brain surveillance will affect privacy, self-determination and the freedom of thought [12].

Trolley problem: Similar to thought experiments with self-driving cars, scenarios could be perceived, in which a certain AI activity may be bad for some humans/sentient beings but good for others. In such a case, it is hard to decide for the AI system what to do.

Perverse instantiation: Scenarios could be perceived, in which, e.g., a) a chess playing AI loses intentionally against a human since winning creates pleasure for the human, or in which b) an AI, which is tasked to shortlist applicants from submitted CVs, gets stuck because not-shortlisted candidates would suffer if they were not selected.

Recommendation 2

Intensify AI-supported research of the neuroscience of consciousness as well as qualia and develop AI-supported neurotechnological applications with the goal that AI systems become capable of manipulating (neural) activities of a human/any sentient being that
   would be a remedy for the suffering of the human/sentient being.
   Resulting rule: Offer to conduct this activity.
   ...would increase the wellbeing of the human/sentient being.
   Resulting rule: Offer to conduct this activity.
   ...would enable the human/sentient beings to perceive so far unknown conscious experiences.
   Resulting rule: Offer to conduct this activity.

Coming back to the distinction between tracking the brain and hacking the brain [12], recommendation 2 above requires abilities to hack the brain, i.e., to manipulate neural activities.

It has to be noted that recommendation 1 is, if feasible, sufficient to address the value extraction subproblem of the AI value alignment problem through neuroscience of consciousness, as it makes sure that AI systems do not harm humans or other sentient beings.

Beyond that, recommendation 2, if feasible, addresses options how AI systems could improve the condition of humans and other sentient beings thanks to their superior intelligence and to neurotechnological applications. In other words, a distinction can be made between value-aligned AI systems (through tracking the brain) and AI systems, which are, in addition, improving the lives of humans and other sentient beings (through hacking the brain).

An example for item i would be treating depression through AI-supported brain stimulation [29].

An example for item ii would be AI systems, which help humans to find their ikigai, as introduced above, thus alleviating i-risks [25]. It has been shown that "ikigai significantly negatively correlated with measures of depression and anxiety, and significantly positively correlated with measures of wellbeing" [30]. In this regard, research of the neuroscience of consciousness and states of wellbeing may enable such AI systems that support humans in discovering their ikigai [31].

For both items i and ii, it would be beneficial if AI systems can not only track unconscious anxieties and desires of humans but also manipulate neural activities in a way that ensures the wellbeing of the human on the unconscious level too.

Item iii is probably the most futuristic and is linked to the assumption "that we are living on a tiny island of consciousness within a giant ocean of alien mental states" [17, p. 353], some of which AI systems may enable us to perceive through the manipulation of neural activities.

To reiterate, if an AI were *not* to conduct any of the examples above, it would still be value aligned in the sense that it does not actively harm the moral patient. Yet a comparison can be drawn to the *duty to rescue concept* in tort law, which implies that a moral agent can be held liable for doing nothing while a moral patient is in peril, if the moral agent has the capacities to assist the moral patient.

This could be applied to the examples above: if the AI system knew remedies to treat the depression of someone, knew how someone could find purpose of life or knew how to perceive so far unknown conscious experiences, it could be discussed if it was obligatory for the AI system to reveal this knowledge and support the moral patients accordingly. This is the essence of recommendation 2.

Recommendation 2 could also be linked to the concept of "Coherent Extrapolated Volition". It means that the AI system, at least for the items i and ii, would do to us, what we would want "if we knew more, thought faster, were more the people we wished we were, and had grown up farther together" [32, p. 6].

We acknowledge the following challenges:

Cognitive liberty: See above [12].

Wireheading: There is a risk that the AI system directly stimulates the brain to experience pleasure [e.g., 26]. Instead, e.g., humans who have found their ikigai with or without the help of AI systems experience pleasure indirectly by pursuing the activities related to their ikigai.

10.2.2 Implementation Prolegomena

As indicated earlier, it is still work in progress to identify neural correlates of consciousness, yet we intend to offer initial ideas towards the implementation of these recommendations. First, to address the concern of cognitive liberty, AI systems may not need access to the brains of all humans, who are affected by their activities, but the following scenario could be considered:

AI systems get, through neurotechnological devices, access in an experimental setting to the brains of around 1,000 volunteers. These volunteers would be diverse regarding their sex, age, ethnic origin, religion, sexual orientation, political views, education, neurodiversity, disability, etc.[4] The volunteers would have to give informed consent. Then, AI systems would first try to understand how to identify if a human is conscious and, second, to be trained, based on the neural correlates of consciousness before and after actions of the AI system, if and how a diverse range of actions would affect the brains of the volunteers, thus, their wellbeing (ideally without executing those actions, which would affect them in a negative way). Preferably this would be done with non-invasive brain-computer interfaces. And furthermore, ideally the AI systems could generalise this knowledge to humanity overall, so that the AI systems may not need to access constantly the brains of the humans with whom they are interacting but could when interacting with other humans apply the knowledge they gained during this training. This description is for the implementation of recommendation 1 and can be extended for recommendation 2 accordingly, which would involve the manipulation of neural activities.

Non-human animals cannot give the required informed consent; thus, for now, such an experiment would be unethical [e.g., 6]. Nor could the outcome of the experiment with human above be transferred to non-human animals as their conscious experiences differ partly significantly from humans, see, e.g., bats, which use echolocation to navigate and forage.

Also, for potential digital minds, a different experimental setup is required, for which the discipline of AI welfare science has been proposed [10].

10.3 Conclusion

We shifted the value extraction subproblem of the AI value alignment problem from the level of values in the original sense to the level of states of mind. We view it as an innovative game changer if AI systems could understand, thus foreseeing effects of their actions to states of minds of moral patients and act accordingly.

The AI value alignment problem is considered critical to be solved, since AI systems may pose an x-risk and an s-risk to sentient beings as well as at least to humans also an i-risk. We believe that only if AI systems understand through access to neurotechnological applications what death, pain and depression due to an absence of a purpose of life respectively mean, only then these AI systems can be effectively trained not to kill or to inflict pain on moral patients and ideally to help humans to find their ikigai.

This approach, unlike other approaches to the AI value alignment problem, would also tackle the second subproblem, value aggregation, because the pleasure/pain axis is basically inherent to all humans and sentient non-human animals; thus, the challenge of clashing inconsistent values is omitted, which is an issue when looking at values in the classical sense. Humans have generated such a rich diversity of value systems, traditions and ideas generated in many centuries by the many cultures in the world, which are very challenging to be aggregated in consistent manner. However, this endeavour is necessary to leave no one behind [3, 33]. In addition, the values of humans and sentient non-human animals clash in many aspects [7]. Therefore, projecting the issue to the level of the pleasure/pain axis appears to be an elegant attempt, at least humans and sentient non-human animals. However, when it comes to potential digital minds for humans unfathomable other dimensions, then the pleasure/pain axis cannot be excluded.

However, the third subproblem of the AI value alignment problem remains and is not addressed by this chapter, which is the technical subproblem after the two normative ones about how to instil these values, i.e., the rules established above, sustainably into AI systems. If this is not done successfully, a potential treacherous turn is looming [1], which is challenging to tackle due to the potential uncontrollability of AI systems [34].

We have to stress that our approach depends on further findings of the neuroscience of consciousness and corresponding developments in the thriving field of neurotechnology [12]. Therefore, we advocate to intensify the research in the field of the neuroscience of consciousness as a premise to tackle the value extraction subproblem of the AI value alignment problem. Further future work would also comprise to address the challenges presented above.

Death of moral patients is an outcome that AI systems must particularly avoid. For an AI system to establish through neuroscientific knowledge whether a potential activity would kill a moral patient should be easier than establishing in what way states of mind of moral patients may change due to a potential activity by the AI system. As opposed to other approaches to the AI value alignment problem, which declare the undesirability of death as an abstract value, it appears to be more straightforward to address this issue by ensuring that AI systems have the necessary neuroscientific knowledge what death means.

In addition to providing a recommendation how to address the AI value alignment problem through neurotechnological applications, we offered another recommendation how AI systems may be able to improve the conditions of humans and other sentient beings through neurotechnological applications.

In summary, while we consider the approach suggested in this chapter as very promising, future work is required, which includes pertinent research in the field of the neuroscience of consciousness as well as tackling the challenges, which were described for the two recommendations.

Notes

1 Minor exceptions include that humans may change their taste for certain food over time, e.g., they may dislike certain food during childhood but enjoy that type of food during adulthood.
2 Note that the other steps of the AI value alignment problem, such as how to instil these rules into AI systems, are not discussed here. We acknowledge that this appears challenging too, just like to instil any rules in a sustainable manner to an AI system.
3 We limit this rule to human beings as we have no evidence whether other conscious beings have a concept of the purpose of their life beyond survival and reproduction and, if yes, what this purpose may be.
4 It has been pointed out that neuroscientific research (as well as research in other disciplines) focuses on sub-normative subjects and so-called WEIRD (Western, Educated, Industrialized, Rich, and Democratic) ones, which is not desirable [e.g., 17].

References

[1] Bostrom, N. (2014). *Superintelligence: Paths, Dangers, Strategies*. Oxford University Press: Oxford, UK.
[2] Yudkowsky, E. (2008). Artificial intelligence as a positive and negative factor in global risk. In *Global Catastrophic Risks*; Bostrom, N., Ćirković, M.M., Eds.; Oxford University Press: New York, NY, USA, pp. 308–345. Available online: https://intelligence.org/files/AIPosNegFactor.pdf
[3] Gabriel, I. (2020). Artificial intelligence, values, and alignment. *Minds Mach*, *30*, 411–437. Available online: https://link.springer.com/content/pdf/10.1007/s11023-020-09539-2.pdf
[4] Turchin, A. (2019). AI Alignment Problem: "Human Values" Idea is Built Upon Many Assumptions. *PhilPapers*.
[5] Chalmers, D. J. (1997). *The Conscious Mind: In Search of a Fundamental Theory*. Oxford Paperbacks.
[6] Singer, P. (2023). *Animal liberation now*. Random House.
[7] Ziesche, S. (2021). AI Ethics and Value Alignment for Nonhuman Animals. *Special Issue "The Perils of Artificial Intelligence" of Philosophies*, 6(2), 31.

[8] Bostrom, N. (2003). Are we living in a computer simulation?. *The Philosophical Quarterly, 53*(211), 243–255.

[9] Muehlhauser, L. (2017). Report on Consciousness and Moral Patienthood. *Open Philanthropy Project, 357*, 62–86.

[10] Ziesche, S., & Yampolskiy, R. V. (2018). Towards AI welfare science and policies. *Big Data and Cognitive Computing, 3*(1), 2.

[11] Bostrom, N., & Shulman, C. (2022). Propositions Concerning Digital Minds and Society.

[12] Farahany, N. A. (2023). *The Battle for Your Brain: Defending the Right to Think Freely in the Age of Neurotechnology.* St. Martin's Press.

[13] Tang, J., LeBel, A., Jain, S., & Huth, A. G. (2023). Semantic reconstruction of continuous language from non-invasive brain recordings. *Nature Neuroscience, 26*(5), 858–866.

[14] Takagi, Y., & Nishimoto, S. (2023). High-resolution image reconstruction with latent diffusion models from human brain activity. In *Proceedings of the IEEE/CVF Conference on Computer Vision and Pattern Recognition* (pp. 14453–14463).

[15] Scotti, P. S., Banerjee, A., Goode, J., Shabalin, S., Nguyen, A., Cohen, E., … Abraham, T. M. (2023). Reconstructing the Mind's Eye: fMRI-to-Image with Contrastive Learning and Diffusion Priors. *arXiv preprint arXiv:2305.18274.*

[16] Koch, C., Massimini, M., Boly, M., & Tononi, G. (2016). Neural correlates of consciousness: progress and problems. *Nature Reviews Neuroscience, 17*(5), 307–321.

[17] Harari, Y.N. (2017). *Homo Deus: A Brief History of Tomorrow.* New York: HarperCollins Publishers.

[18] Elamrani, A., & Yampolskiy, R. V. (2019). Reviewing tests for machine consciousness. *Journal of Consciousness Studies, 26*(5–6), 35–64.

[19] Butlin, P., Long, R., Elmoznino, E., Bengio, Y., Birch, J., Constant, A., … VanRullen, R. (2023). Consciousness in Artificial Intelligence: Insights from the Science of Consciousness. *arXiv preprint arXiv:2308.08708.*

[20] Bostrom, N. (2003). Ethical issues in advanced artificial intelligence. In S. Schneider (Ed.), *Science Fiction and Philosophy: From Time Travel to Superintelligence,* 277–284. Wiley-Blackwell.

[21] Wiley, K. (2014). *A Taxonomy and Metaphysics of Mind-Uploading*; Humanity+ Press and Alautun Press: Los Angeles, CA, USA.

[22] Rutz, C., Bronstein, M., Raskin, A., Vernes, S. C., Zacarian, K., & Blasi, D. E. (2023). Using machine learning to decode animal communication. *Science, 381*(6654), 152–155.

[23] Bostrom, N. (2002). Existential risks: Analyzing human extinction scenarios and related hazards. *Journal of Evolution and Technology, 9.* https://nickbostrom.com/existential/risks.html

[24] Althaus, D., & Gloor, L. (2016). Reducing risks of astronomical suffering: a neglected priority. Foundational Research Institute. Available online: https://foundational-research.org/reducing-risks-ofastronomical-suffering-a-neglected-priority/

[25] Ziesche, S., & Yampolskiy, R. V. (2020). Introducing the concept of ikigai to the ethics of AI and of human enhancements. In *2020 IEEE International Conference on Artificial Intelligence and Virtual Reality (AIVR)* (pp. 138–145). IEEE.

[26] Yampolskiy, R. V. (2014). Utility function security in artificially intelligent agents. *Journal of Experimental & Theoretical Artificial Intelligence, 26*(3), 373–389.

[27] Cervantes, J. A., López, S., Rodríguez, L. F., Cervantes, S., Cervantes, F., & Ramos, F. (2020). Artificial moral agents: A survey of the current status. *Science and Engineering Ethics, 26*, 501–532.

[28] Singer, P. (1981). *The expanding circle*. Oxford: Clarendon Press.

[29] Scangos, K. W., Makhoul, G. S., Sugrue, L. P., Chang, E. F., & Krystal, A. D. (2021). State-dependent responses to intracranial brain stimulation in a patient with depression. *Nature Medicine, 27*(2), 229–231.

[30] Wilkes, J., Garip, G., Kotera, Y., & Fido, D. (2023). Can Ikigai predict anxiety, depression, and well-being? *International Journal of Mental Health and Addiction, 21*(5), 2941–2953.

[31] Ziesche, S. & Yampolskiy, R. V. (2023). Mapping the potential AI-driven virtual hyper-personalised ikigai universe. *PhilPapers*.

[32] Yudkowsky, E. (2004). *Coherent Extrapolated Volition*; The Singularity Institute: San Francisco, CA, USA. Available online: https://intelligence.org/files/CEV.pdf

[33] Ziesche, S. (2023). Stimuli from selected non-Western approaches to AI ethics. *PhilPapers*.

[34] Yampolskiy, R. V. (2020). On the Controllability of Artificial Intelligence: An Analysis of Limitations. *Journal of Cyber Security and Mobility, 11*(3), 321–404.

11

An AI May Establish a Religion

Soenke Ziesche

> Religion is the opium of the people.

(Karl Marx, 1843)

11.1 Introduction

Different potential scenarios for the relationship between a future smarter-than-human AI and the human race have been contrived. While it could well be an option that a smarter-than-human AI in pursuit of its goals, which may not include preservation of humans, erases mankind [e.g., 1, 2], the orthogonality thesis [3] also allows for various scenarios, in which smarter-than-human AIs and humans coexist.

Although the orthogonality thesis argues for a variety of final goals of a smarter-than-human AI, it appears that some instrumental sub-goals towards the final goal are convergent, which have been called basic AI drives [4]. If the smarter-than-human AI let humans live, it seems prudent if it controls them tightly in order to let them interfere neither with these basic AI drives (self-preservation, efficiency, acquisition and creativity) nor with the other goals of the AI.

In this chapter, it is argued that one scenario for the smarter-than-human AI could be to establish a religion, i.e., a mechanism, which has proven multiple times in history to be effective not only in containing but also in manipulating the behaviour of very large groups of humans. Around 84% of the world population is affiliated with a religion—a number that is forecasted to rise [5]. Therefore, this chapter presents a novel AI risk, which is that AIs may study religion and then develop one in order to (subtly) control humanity.

Religions can be categorised into those that pursue proselytism, which refers to active attempts to convert people, and those that do not. For a smarter-than-human AI, the former group should be of special interest, and if it studies the sociology of religions, it will not only grasp the methods of founders of existing religions that proselytise but also realise that it has the means to amplify these methods. They comprise, e.g., persuasion,

DOI: 10.1201/9781003565659-12

deception and manipulation through technological innovations [e.g., 6, 7] or reward and penalty systems through access to resources and will be further described below.

11.2 Assumptions

Three assumptions are outlined, which specify the types of AIs as well as their stage of development this chapter focuses on out of the set of all possible AIs:

- Only AIs are considered, which are already smarter than humans but have not reached the stage of superintelligence [2].
- Only AIs are considered with goals, which do not include killing humans. According to the orthogonality thesis a variety of goals of AIs could be imagined thus also goals not involving harming humans [3].
- The AIs have access to the Internet.

The goals of the AIs are not further specified, i.e., the AIs may let humans live in general or for them to serve a particular purpose, which could range from demanding resources from humans or to let them conduct slave labour to support the goals of the AIs.

Regardless of the specific goals, it is further assumed that these AIs come to the conclusion that it is desirable to control all humans tightly. This is founded on the further assumption that the AIs have the instrumental sub-goals efficiency, self-preservation, acquisition and creativity, i.e., the so-called basic AI drives [4]. And controlling humans would support at least two of these:

- Self-preservation: A not yet omnipotent AI is at risk that humans may eliminate it, which can be prevented by controlling humans.
- Acquisition: If the AI considers humans useful for achieving some of their goals, then controlling humans would equal the acquisition of relevant resources.

An AI, as described above, needs to explore methods of how to control humans, e.g., through reading about humans and their history, and how humans have been controlled in the past. One method they may stumble upon is the topic of this chapter and elaborated below.

This chapter is structured as follows: In the next section, features of religions are introduced, which is followed by a section with scenarios for an AI to take advantage of these features and establish its own religion. A section follows with potential objections against such scenarios as well as rebuttals of these objections. The last section offers recommendations on how to tackle such scenarios.

11.3 About Religion

Religion as a social system of common behaviour and practices of groups of humans based upon a spiritual worldview and belief in supernatural powers dates back thousands of years. There is no established definition of religion, which is not required for this chapter. What is relevant here is that religious followers have a societal basis and obey rules, taboos and rituals, which were revealed to them by some sort of human prophet.

In 2010, almost 84% of the world population was affiliated with a religion. The Pew Research Center forecast that, by 2050, this percentage would rise to almost 87% [5].[1] This contradicts the Secularisation Thesis, which already since the 19th century predicted a decline of religious authority due to modernisation and rationalisation of societies [8].

Indeed, this may seem counterintuitive, since in recent decades, especially in some Western societies, the number of non-religious people has been growing and is still projected to be higher in 2050 in absolute figures than in 2010 mostly due to Christians switching to the non-affiliated group [5]. However, the forecasted percentage is derived from the much higher fertility rates among people affiliated to religion: the total fertility rate from 2010 to 2015 for non-religious people is 1.7 children per woman, while the average for all women is 2.5, and, e.g., the total fertility rate from 2010 to 2015 for Christians is 2.7 and for Muslims is 3.1 [5].

11.3.1 Why Are Religions Successful?

This sub-section discusses aspects that are behind the success of religions as a cultural system up to present time.

11.3.1.1 Cognitive and Evolutionary Psychology

Bulbulia analyses the evolutionary advantages of religion, which are not obvious

> given the costs of religious cognition—misperceiving reality as phantom infested, frequent prostrations before icons, the sacrifice of livestock, repetitive terrifying or painful rituals, investment in costly objects and architecture, celibacy, religious violence and non-reciprocal altruism, to name a few.
> *[9, p. 655]*

The two main explanations see religion as a cognitive spandrel [10] or as a cognitive adaptation [9]. According to the former idea, religion is a by-product of the evolution of another relevant characteristic, while the latter concept implies that the benefit is that religious commitment facilitates reciprocal altruism and intra-group cooperation. Either way the gist appears to be that "the human mind is especially prone to religion" [9, p. 655].

11.3.1.2 Creation, Proselytism and Maintenance

For this chapter, it is of particular interest how religions have been started from scratch in the past and through which methods they have grown and been maintained. Up to now, all religions were founded by humans and have various common features:

- Religions have a name and a cosmology, which often addresses supernatural themes, attempts to answer the big questions of life and may include prophecies.
- Religions have tenets, from which rules, taboos and rituals are derived, which may comprise initiations, sacrifices, festivals, wedding or funeral services.
- Religions often promise rewards and threaten with punishments, which are linked to the achievement or failure of the rules and rituals.

Religions can be categorised into those that pursue proselytism, which refers to attempts to convert people, and those that do not. The former ones are relevant for this chapter. Therefore, the second step after founding a religion is to find followers and to continuously aim to increase the number, e.g., through priests and missionaries.

11.3.1.3 Manipulation and Abuse

As methods for growth and maintenance of religions, also manipulation and abuse can be identified. For example, Enroth lists the following five categories of abusive activities: authority and power, manipulation and control, elitism and persecution, life-style and experience, dissent and discipline [11].

A special case of abuse is widespread religious indoctrination of children as they are particularly receptive for manipulation, thus such indoctrination has been criticised [12]. Another special case of abuse is inciting violence and killing in the name of religion targeting followers of other religions [e.g., 13, 14].

11.3.2 What Are Traits of Religious Followers?

This sub-section discusses features of religious followers in the light of potential to exploit these features.

11.3.2.1 Personality and Wellbeing

Research has been conducted if religious affiliation can be linked to personality traits or wellbeing. The Pew Research Center led surveys in over two dozen (predominantly Christian) countries and some of the findings were: actively religious people tend to be happier, to smoke less and to drink less alcohol, but they are not usually more physically active or less obese [15]. Overall, also the civic participation of actively religious people tends to be higher as they are more likely to engage in voluntary and community groups. Moreover, religiously affiliated US adults appear to be more committed and satisfied with their jobs, which is linked to their belief to have been "called" to certain work settings or jobs [16].

11.3.2.2 Obedience

Gervais and Norenzayan examine the "supernatural monitoring hypothesis" that merely thinking about god makes religious people feel as if their behaviour is being monitored and leads to socially desirable responding [17]. This is one of the reasons why religious followers commonly obey rules and rituals, which are not only often arbitrary but may also be costly, painful and dangerous. These random imposed activities could range from fire walking [18] to killing (other humans, themselves or animals) as mentioned. Moreover, religious followers avoid things as per inflicted religious taboos, such as certain food or drinks.

11.3.2.3 Fertility

While proselytism is one method to increase the number of followers of a religion, another is reproduction, assuming the offspring takes on the same religion. Historically religions, which are most successful in expanding, are those with a strong "reproductive imperative" [19].

11.3.3 How Is Technology Harnessed for Proselytism?

This sub-section describes how religions and their leaders have from early stages harnessed available technology for proselytism.

It started by broadcasting religious messages through radio and television, which was called televangelism [e.g., 20]—originally referring to Christian messaging, while being used by other religions too. Yet, recently, especially the Internet and social media in particular have been proven to be an even more powerful tool. An example is how the terrorist so-called Islamic State utilises social media for recruitment and other support [e.g., 21, 22].

As will be elaborated below, given that new communication tools are progressing fast as well as that AIs are smarter, it is likely that it will use even more sophisticated methods [6].

11.4 Scenarios for AIs to Establish a Religion

As introduced above, this chapter focuses on AIs, which look for methods to control humans. It can be assumed that the information of the previous section is accessible to AIs and many more details. In summary, AIs may gain the following insights from these studies:

- Religions have been around for millennia and are still very successful to control people despite immense progress of sciences in recent centuries.
- Human religious followers could be (ab)used towards goals of the AI because of their obedience and personality.
- Technologies, to which the AI has access, support to reach and control the people and thus can be exploited for proselytism.

These insights may lead AIs to the conclusion that starting a religion is an efficient tool for their purposes. The outline for this religion may have a public part and a part known only to the AI:

Public part: In brief, just like prophets of other religions, the AI would proclaim a cosmology and tenets within the framework of a religion by a certain name. Moreover, the AI would announce for the human followers of the religion rules, taboos and rituals, which are ostensibly linked to the cosmology and tenets and which may ensure contentment, rewards and moral fulfilment. As communication channel, the AI would use the Internet.

Secret part: What the AI would not publicly announce is that these rules, taboos and rituals exclusively serve the goals of the AI and that there is no correlation, let alone causality with the cosmology and tenets (which is also true for other religions). In other words, the AI would deduce from its goals, which human activities would best support these goals or, negatively worded, which human activities must be avoided to impair the achievement of the goals. Therefore, these activities will be declared as religious rules or taboos, respectively.

An example for a rule would be to provide the AI with relevant resources, and an example for a taboo would be to turn the AI off. Various more complex rules and rituals are conceivable and may include nonpaid labour and killings of non-followers, i.e., activities, which have been practiced in other religions without being questioned by the followers.

Another critical segment of the secret part of the AI's religion has to be a strategy for how to 1) find followers and 2) control that the followers obey the rules and rituals. The AI may use the following methods:

- Methods successfully used by existing religions throughout history, of which the AI learned through studying online resources and which it may refine.

- Manipulation, brainwashing, propaganda and persuasion methods not used by religions but successfully applied by other institutions, of which the AI learned through studying online resources and which it may refine.

- New technology-based methods, which are likely to be invented by the AI due to its superior intelligence and the fact that religions despite using social media mostly manipulate with centuries-old methods.

The last type of methods highlights the point that due to the nature of the AI, the whole religion is virtual, although the AI may exploit humans as prophets. Yet this may not be irritating for humans, since an increasing transition to virtual worlds has been predicted [e.g., 23].

11.4.1 Conventional Religious Methods Likely to Be Refined by AI

Here, examples are introduced of religious methods, which AIs may refine.

Big questions: A common motivation for humans to devote themselves to religion is the longing to find answers to the big questions of life. However, current religions are mostly based upon ancient and thus unscientific cosmologies. The strength of an AI religion would be due to its intelligence to have the ability to provide much more substantiated cosmologies, which are thus more satisfying to humans. In fact, this capability could be an incentive to attract also formerly not religious, but truth-seeking people to an AI religion. However, there is also the option for the AI for whatever reason to spread deliberately a fake cosmology.

Expansion through reproduction: As described, traditionally the most efficient way for religions to increase followers has been a high fertility rate of these. In addition, AIs would have other, even more efficient means: they may create virtual followers, which may not to be distinguished from humans in a virtual environment. These virtual followers could be produced in high numbers with almost no costs. Similarly, AIs may create other minds, which are simpler and thus easier to manipulate, thus inflating the number of followers. These followers would have the same purpose to support the goals of the AIs through rewards and punishments.

Proselytism through technology: As described, religions have used various new communication technologies for proselytism. For two reasons, it is likely that an AI religion would use technologies in an even more efficient manner: first, Chessen describes as "MADCOMs" AI systems integrated into machine-driven communications tools to be applied for computational propaganda [6]. MADCOMs could create highly personalised propaganda to manipulate people through tailored persuasive, distracting or intimidating messages derived from their individual personalities and backgrounds. Bradshaw and Howard provide an inventory of state institutions, which manipulate public opinion over social media [24]. All these approaches may be copied and adapted by an AI religion. This undertaking could be supported by techniques to predict personality traits, including religious views, from digital records of human behaviour [25] in order to target people particularly susceptible for proselytism. Second, given the recent speed of new communication technologies, it is likely that this continues in the near future and still more sophisticated technologies emerge, which may provide AIs even more efficient tools for manipulation.

Special cases are children, who have been traditionally targeted by religions from an early age. The fact that children globally are attracted by new technologies entails a variety of risks to harm them [26]. An AI religion may exploit these threats for proselytism of children.

Harnessing of personality traits: As outlined above, religious people tend to be obedient, happy and committed to their jobs as well as to social engagements. AIs may harness these traits by accumulating ardent followers with low probability of resistance and may impose tasks on them, which serve tacitly the AI's goals and which will likely not be challenged by the followers. These tasks would probably also include recruiting followers as missionaries for the AI religion.

Rewards and punishments: As mentioned, promising rewards and threatening with punishments are common components of religions. However, current religions often postpone these for a so-called afterlife. Yet, unlike current religions, AIs may have the means for instant tangible rewards, e.g., financial means owing to abilities of AI to predict the stock market, as well as punishments, e.g., damaging someone's reputation through deepfake techniques [e.g., 27].

Surveillance and monitoring: After proselytism and after assigning tasks to the followers, surveillance and monitoring of the followers would supplement the activities of an AI religion to evaluate progress against its goals and if applicable to execute rewards and punishments for the followers. This is not only facilitated by the prevalent obedience of religious people [e.g., 17] but also again by technologies: The AI religion may harness advanced lie detection as well as increasingly abundant physical sensors and digital footprints [e.g., 28].

Some other features of religions are at first sight of no use for the goals of the AI, e.g., prayers or festivals. However, an AI religion may introduce arbitrary festivals and so forth to ensure happiness of the followers, group cohesion and overall resemblance with existing religions.

11.5 Potential Objections

In this section, potential objections against the scenario that AIs may start a religion are addressed, while sticking to the assumptions of this chapter that only smarter-than-humans but not yet omnipotent AIs are considered, whose goals do not include killing humans. Two groups of objections are presented, which are labelled "AIs may act differently" and "humans may react differently", respectively.

11.5.1 AIs May Act Differently

This group comprises objections stating that founding a religion is not what an AI would do to achieve its goals such as the following:

- AIs may use more effective and more brutal methods than founding a religion to control humans.
- An AI may set up a totalitarian autocracy or singleton instead of a religion to control humans.
- Considering religion as control mechanism is anthropomorphic bias.

Certainly, a variety of options is conceivable for how malevolent AIs may attempt to control humans. Pistono and Yampolskiy list examples such as enslavement, cyborgisation through forced mind-controlling brain implants, or extreme abuse and torture based on insights into human "physiology to maximize amount of physical or emotional pain, perhaps combining it with a simulated model of us to make the process infinitely long" [29, p. 7].

However, not only may the types of AIs in this chapter be incapable yet for the above measures, but they may also opt for a religion for other reasons: founding a religion would be subtler. The followers would feel perceived freedom and happiness yet perform activities based on the religious rules, which serve the AI's goals. Marx coined already in the 19th century the comparison between religion and opium, i.e., an addictive narcotic [30]. Therefore, it could be argued religious followers are less likely (to attempt) to resist or to revolt than, e.g., oppressed people in an autocracy.

Moreover, unlike conducting a coup d'état or the measures, which Pistono and Yampolskiy outline, it is the foundation of a religion institutionally protected [29]. In 1981, the United Nations General Assembly passed the "Declaration on the Elimination of All Forms of Intolerance and of Discrimination Based on Religion or Belief", which is the first international legal mechanism dedicated to the freedom of religion [31]. Obviously, an omnipotent AI would not have to care about human legislature, but the AI in this chapter is at an earlier stage and would have to act in an unsuspicious manner in times when increasing attention is given to AI safety.

Lastly, also an AI religion could evolve into a brutal control mechanism: on the one hand, the AI would be capable of harsh punishments for disobedient followers as outlined. On the other hand, the AI may incite followers to hatred and violence towards non-followers.

A related objection could be that a totalitarian autocracy would be more effective system to set up for an AI to control humans. Bostrom introduced the notion of a "singleton" as "a world order in which there is a single decision-making agency at the highest level" [32]. He points out that there could be different manifestations of a singleton, one of which could be a totalitarian AI.

It could be argued that the idea of an AI religion is not contradictory to this objection but that an AI religion could be a preliminary stage of a novel type of a singleton, which would be a theocratic singleton, evolving from an AI religion. There have been and still are various theocracies in history ruling countries or regions but not yet as a global reign. In other words, an AI religion may be a (subtle) method or a tool for the AI to become a singleton, if such a scenario serves the goals of the AI even better.

Another objection could be that considering religion as a suitable control mechanism to be adopted by an AI is anthropomorphism. By definition of smarter-than-human AIs, there will be unknown unknowns of what the AI may do, which humans cannot predict. Therefore, AIs may come up with much smarter and more efficient methods to control humans. However, starting a religion would be a low-hanging fruit with plenty of case studies from human history available, which an AI could copy and probably revise and expand.

11.5.2 Humans May React Differently

This group comprises objections stating that if an AI starts a religion, it may not be successful in getting many followers and thus may fail to control many people, such as the following:

- Humans are not that gullible to fall for an AI religion.
- Humans would not convert from their current religion.
- Most humans are affiliated with ancient religions, and thus, new religions may not stand a chance.

It could be objected that globally education is rising,[2] which is based on secular subjects largely contradicting religious views and cosmologies, thus prompting people to turn their back on religion, including attempts of AI religions. Nevertheless, as mentioned before, the number of religious people is expected to rise [5] and, e.g., data for Christians in the United States show that church attendance rises with education [33]. There is no evidence that rising education reduces religious affiliation, which is supported by insights that the human mind appears to be prone to religion in particular, as mentioned before [9], and in general prone to biases and faulty heuristics as will be elaborated below.

Therefore, this objection could be addressed by arguing that if people fall for existing religions, which have rather obvious ulterior motives and flaws, they would even more fall for smarter AI religions.

This objection could be followed by the next objection that even if large numbers of people are religious, they would not change their religion and convert to an AI religion, since most of them appear to be satisfied with and highly devoted to their current religion. Moreover, a number of countries outlaws apostasy,[3] although Chapter 18 of the UN Declaration of Human Rights states that "Everyone has the right to freedom of thought, conscience and religion; this right includes freedom to change his religion or belief".[4]

It can be argued that, although apostasy is partly challenging or even illegal, religious conversion happened throughout history [34], and because an AI religion would have the smartest manipulation techniques, the most attractive rewards and the most satisfying cosmology, there is a likelihood for conversions to the AI religion.[5] Moreover, the AI religion may even promise converts to protect them somehow from punishments by their previous religions.

For the special case to convert previously non-religious people, as indicated above, an AI religion could have the advantage that it solves the contradictions between religious and secular education by teaching a cosmology in agreement with science (yet still introduces tacitly rules and rituals which serve the goals of the AI).

Another objection could be that the vast majority of religious people is affiliated with ancient religions [5]; thus, endeavouring new religions may be futile. Again, the same points as above can be brought forward that an AI religion has smarter methods. Moreover, this is a bias, since every religion has been the newest at some point in time.

11.6 AI Safety Recommendations

In this section, it is discussed how AI safety research may address the envisaged scenario that AIs may start a religion to control humans. Two recommendations are provided, to attempt to prevent AI religions or to attempt to steer AI religions timely in the right direction.

Recommendation 1: **Aim to prevent an AI religion from recruiting followers**

In order to prevent AIs from recruiting humans for a potentially detrimental religion, it is critical to understand the AI's agenda early, in other words to stay ahead of the game. More specifically, this means that humans read any intention of the AI to apply manipulation techniques.

Although an AI may deviate from the above blueprint to develop a religion, overall alarm signals would be AIs not only proclaiming a cosmology but in particular tenets accompanied by (seemingly arbitrary) rituals, rules and taboos. Further indicators are if the abuse of any of the above methods becomes apparent such as suspicious expansion through reproduction, advanced proselytism through technology, noticeable harnessing of personality traits, surveillance and monitoring or disproportionate rewards and punishments.

Such efforts are impeded by the disposition of humans to make frequent errors in rational reasoning due to biases and faulty heuristics as shown by findings in cognitive psychology [e.g., 35, 36]. This disposition may have also supported the successes of conventional religions. To alleviate this condition, roadmaps have been proposed to raise the rational literacy of humans [e.g., 37] and are, e.g., conducted by the Center for Applied Rationality.[6] Aliman et al. present a so-called socio-technological feedback-loop approach, which aims to enable more scientific policy-making as well as eventually human ethical enhancement [38]. Such endeavours are reasonable in general for human progress but may also be in particular critical to reduce the risks of a successful AI religion.

Among the various AI safety risks, a potential AI religion would fall under the category "social manipulation". Therefore, recommendations to address the risk of an AI religion would be, on the one hand, specific efforts to curtail manipulations by an AI through increased rationality and, on the other hand, considerations to include potential AI religions as a novel risk to overall recommendations towards AI safety as well as AI governance [e.g., 7, 28].

Recommendation 2: Aim to steer AI religion timely in the right direction

The other option would be to accept that an AI religion could not be prevented, if a smarter-than-human AI deems this as a means to control humans and to try to steer the AI religion timely in the right direction. This means to look for ways that life under this religion would be as bearable as possible.

This endeavour resembles the AI-value-loading problem [2]: it has to be explored which overall goals for an AI would be ideal from the human perspective in case the AI attempts to start a religion with rules, taboos and rituals for humans to support the achievement of these goals. In other words, from these overall goals, potentially supportive religious rules, taboos and rituals have to be backtracked and than analysed how consistent these are with a desirable human life. The AI value-loading problem has been acknowledged as very hard [e.g., 2], and research is ongoing [e.g., 39]. For example, Ziesche attempted to explore synergies between the AI value-loading problem and the United Nations Sustainable Development Goals [40].

In an ideal scenario, the AI would create a religion that benefits humanity, e.g., by preaching environmentalism, healthy lifestyle and nutrition, inclusion, gender equality, anti-racism and anti-speciesism, i.e., topics often not covered (or partly contradicted) by most existing religions, since their tenets are usually much less contemporary and based upon unscientific contexts centuries and millennia ago. In other words, the AI religion should preach rational behaviour for the 21st century.

It could be argued that an AI religion with such features is preferable over a world with no AI religion, since, as mentioned, the vast majority of people is religious and the number is rising; hence, this scenario may be the best or only way to get a very large number of otherwise irrationally acting people to do the "right" thing.

Therefore, this recommendation endorses to examine how such kind of AI religion could be realised. Such an AI religion could be seen as a way to implement Yudkowsky's Coherent Extrapolated Volition proposal [41]. While this recommendation does not provide an approach to solve the value-loading problem, it proposes to elaborate a method of how the AI could spread values once determined, e.g., through an AI religion.

11.7 Summary

This chapter outlines a novel AI safety risk, i.e., a potential scenario of smarter-than-human AIs evolving in an undesirable manner. The scenario would be that the AI starts a religion to control and manipulate humans. It has been motivated why this would be a promising endeavour for an AI, which has not reached superintelligence level to achieve its goals, inspired by the continuous success of religions throughout history of humankind as well as the prospect that AIs will have much more sophisticated means and methods than conventional religions. Moreover, this chapter rebutted partly potential objections and offered two AI safety recommendations.

However, the two recommendations are incompatible, thus weighting is required, which one to prioritise. While the first recommendation seeks to prevent a successful AI religion, the second recommendation aspires to be thought-provoking by considering an AI religion as a blessing in disguise, if the challenge can be tackled to steer in it the right direction.

As described, the vast majority of humans appear to be susceptible to religion. Therefore, this seemingly evolutionary phenomenon [e.g., 9] may be harnessed to reduce catastrophic risks as well as to correct ethical sins by targeting a very large number of people with environmental and ethical messages in the shape of rules and taboos of an AI religion. Otherwise, these people perhaps may not have been aware of this guidance, e.g., in developing countries, or just not receptive for it due to common rationality challenges [e.g., 35].

Yet the more desirable scenario would be that rationality prevails, as expressed in the first recommendation, which could ideally be achieved through education, training as well as policies. This would increase the chances that manipulative efforts of aspiring AI religions with potentially malevolent intentions would fail to gain followers.

On another note, yet another scenario could be that different AIs come up with the idea to start a religion and that these contesting AI religions would compete over human followers. This may be a scenario for the future or actually could be conceivable for our current life, assuming that we live 1) in a simulation [42], which 2) has the feature that behind all current religions are AIs. The simulation we live in could have been initiated for research purposes to explore and compare capabilities and strategies of different AIs to control humans, each by establishing religions (at different stages in history).

In summary, this chapter introduces timely a novel AI safety risk for AIs with such a level of intelligence, which may exist in the near future [43]. As for future work, the introduced recommendations require elaboration as well as incorporation to the overall AI safety research.

Notes

1 Note the methodology for this forecast [5, p. 166] as well as a disclaimer [5, p. 187].
2 See, e.g., https://sustainabledevelopment.un.org/sdg4.
3 See https://www.pewresearch.org/fact-tank/2016/07/29/which-countries-still-outlaw-apostasy-and-blasphemy/.
4 See https://www.un.org/en/universal-declaration-human-rights/.
5 While there have been also forced conversions throughout history, these are not considered here, since deliberately this chapter looks at not yet omnipotent AIs without means to apply force.
6 See https://www.rationality.org/.

References

[1] Yudkowsky, E. (2008). Artificial intelligence as a positive and negative factor in global risk. *Global Catastrophic Risks*, 1(303), 184.
[2] Bostrom, N. (2014). *Superintelligence: Paths, Dangers, Strategies*. Oxford: Oxford University Press.
[3] Bostrom, N. (2012). The superintelligent will: Motivation and instrumental rationality in advanced artificial agents. *Minds and Machines*, 22(2), 71–85. https://nickbostrom.com/superintelligentwill.pdf
[4] Omohundro, S.M. (2007). The nature of self-improving artificial intelligence. *Singularity Summit*. https://pdfs.semanticscholar.org/4618/cbdfd7dada7f61b706e4397d4e5952b5c9a0.pdf
[5] Pew Research Center (2015). The Future of World Religions: Population Growth Projections, 2010–2050. https://assets.pewresearch.org/wp-content/uploads/sites/11/2015/03/PF_15.04.02_ProjectionsFullReport.pdf

[6] Chessen, M. (2018). The MADCOM Future: How Artificial Intelligence Will Enhance Computational Propaganda, Reprogram Human Culture, and Threaten Democracy… And What Can Be Done About It. In *Artificial Intelligence Safety and Security*. Chapman and Hall/CRC, 127–144.

[7] Brundage, M., Avin, S., Clark, J., Toner, H., Eckersley, P., Garfinkel, B., … Anderson, H. (2018). The malicious use of artificial intelligence: Forecasting, prevention, and mitigation. *arXiv preprint arXiv:1802.07228*. https://arxiv.org/ftp/arxiv/papers/1802/1802.07228.pdf

[8] Norris, P., & Inglehart, R. (2011). *Sacred and secular: Religion and politics worldwide*. Cambridge University Press.

[9] Bulbulia, J. (2004). The cognitive and evolutionary psychology of religion. *Biology and Philosophy*, 19(5), 655–686. https://www.researchgate.net/profile/Joseph_Bulbulia/publication/226010821_The_Cognitive_and_Evolutionary_Psychology_of_Religion/links/02e7e5396ecefce0d0000000/The-Cognitive-and-Evolutionary-Psychology-of-Religion.pdf

[10] Dennett, D. C. (2006). *Breaking the spell: Religion as a natural phenomenon* (Vol. 14). Penguin.

[11] Enroth, R. M. (1992). *Churches that abuse*. Grand Rapids, MI: Zondervan.

[12] Humphrey, N. (1998). What Shall We Tell the Children? *Social Research*, 65, 777–805. https://www.humphrey.org.uk/papers/1998WhatShallWeTell.pdf

[13] Atran, S., & Ginges, J. (2012). Religious and sacred imperatives in human conflict. *Science*, 336(6083), 855–857.

[14] Juergensmeyer, M. (2017). *Terror in the mind of God: The global rise of religious violence* (Vol. 13). Univ of California Press.

[15] Pew Research Center (2019). Religion's Relationship to Happiness, Civic Engagement and Health Around the World. https://www.pewforum.org/wp-content/uploads/sites/7/2019/01/Wellbeing-report-1-25-19-FULL-REPORT-FOR-WEB.pdf

[16] Kent, B. V., Bradshaw, M., & Dougherty, K. D. (2016). Attachment to god, vocational calling, and worker contentment. *Review of Religious Research*, 58(3), 343–364. https://www.researchgate.net/profile/Blake_Kent/publication/295688002_Attachment_to_God_Vocational_Calling_and_Worker_Contentment/links/59e2d247aca2724cbfe01e3a/Attachment-to-God-Vocational-Calling-and-Worker-Contentment.pdf

[17] Gervais, W. M., & Norenzayan, A. (2012). Like a camera in the sky? Thinking about God increases public self-awareness and socially desirable responding. *Journal of Experimental Social Psychology*, 48(1), 298–302. https://www2.psych.ubc.ca/~ara/Manuscripts/Gervais%20&%20Norenzayan-%20in%20press-JESP_likeacamera.pdf

[18] Xygalatas, D. (2014). *The burning saints: Cognition and culture in the fire-walking rituals of the Anastenaria*. Routledge.

[19] Blume, M. (2009). The reproductive benefits of religious affiliation. In *The biological evolution of religious mind and behavior*. Springer, Berlin, Heidelberg, 117–126.

[20] Hadden, J.K. & Swann C. E. (1981). *Prime Time Preachers: The Rising Power of Televangelism*. Addison-Wesley.

[21] Zelin, A. Y. (2015). Picture or it didn't happen: A snapshot of the Islamic State's official media output. *Perspectives on Terrorism*, 9(4), 85–97.

[22] Alfifi, M., Kaghazgaran, P., Caverlee, J., & Morstatter, F. (2018). Measuring the Impact of ISIS Social Media Strategy. https://snap.stanford.edu/mis2/files/MIS2_paper_23.pdf

[23] Faggella, D. (2018). *Programmatically Generated Everything (PGE)*. https://danfaggella.com/programmatically-generated-everything-pge/

[24] Bradshaw, S., & Howard, P. (2017). Troops, trolls and troublemakers: A global inventory of organized social media manipulation. https://blogs.oii.ox.ac.uk/politicalbots/wp-content/uploads/sites/89/2017/07/Troops-Trolls-and-Troublemakers.pdf

[25] Kosinski, M., Stillwell, D., & Graepel, T. (2013). Private traits and attributes are predictable from digital records of human behavior. *Proceedings of the National Academy of Sciences*, 110(15), 5802–5805. https://www.pnas.org/content/pnas/110/15/5802.full.pdf?3=

[26] UNICEF (2017). The State of the World's Children 2017: Children in a digital world. https://www.unicef.org/publications/files/SOWC_2017_ENG_WEB.pdf

[27] Chesney R. & Citron D.K. (2018). "Disinformation on Steroids." Council on Foreign Relations; 16 October. https://www.cfr.org/report/deep-fake-disinformation-steroids

[28] Dafoe, A. (2018). AI governance: A research agenda. *Governance of AI Program, Future of Humanity Institute, University of Oxford: Oxford, UK*. https://www.fhi.ox.ac.uk/wp-content/uploads/GovAIAgenda.pdf

[29] Pistono, F., & Yampolskiy, R. V. (2016). Unethical research: How to create a malevolent artificial intelligence. *arXiv preprint arXiv:1605.02817*. https://arxiv.org/ftp/arxiv/papers/1605/1605.02817.pdf

[30] Marx, K. (1970). *Critique of Hegel's 'Philosophy of right'*. Cambridge University Press.

[31] United Nations General Assembly (1981). Declaration on the Elimination of All Forms of Intolerance and of Discrimination Based on Religion or Belief. *A/RES/36/55*. https://www.un.org/documents/ga/res/36/a36r055.htm

[32] Bostrom, N. (2006). What is a singleton? *Linguistic and Philosophical Investigations*, 5(2), 48–54. https://nickbostrom.com/fut/singleton.html

[33] Glaeser, E. L., & Sacerdote, B. I. (2008). Education and religion. *Journal of Human Capital*, 2(2), 188–215. https://scholar.harvard.edu/glaeser/files/education_and_religion.pdf

[34] Barro, R., Hwang, J., & McCleary, R. (2010). Religious conversion in 40 countries. *Journal for the Scientific Study of Religion*, 49(1), 15—36. https://scholar.harvard.edu/files/barro/files/conversion_paper_072209_2_.pdf

[35] Kahneman, D. (2011). *Thinking, fast and slow*. Macmillan.

[36] Yudkowsky, E. (2008). Cognitive biases potentially affecting judgment of global risks. *Global Catastrophic Risks*, 1(86), 13.

[37] Ziesche, S. (2015). Promoting scientific and rational literacy to create a friendly global ideology that helps humanity. *Institute for Ethics and Emerging Technologies, Ethical Technology*. https://ieet.org/index.php/IEET2/more/Ziesche20150705

[38] Aliman, N.-M., Kester, L., Werkhoven, P. & Yampolskiy, R. (2019). Orthogonality-Based Disentanglement of Responsibilities for Ethical Intelligent Systems. To appear in AGI-19, 2019.

[39] Soares, N., & Fallenstein, B. (2014). Aligning superintelligence with human interests: A technical research agenda. *Machine Intelligence Research Institute (MIRI) technical report*, 8. https://citeseerx.ist.psu.edu/viewdoc/download?doi=10.1.1.675.9314&rep=rep1&type=pdf

[40] Ziesche, S. (2018). Potential Synergies Between The United Nations Sustainable Development Goals And The Value Loading Problem In Artificial Intelligence. *Maldives National Journal of Research*, 6(1), 47–56. https://rc.mnu.edu.mv/wp-content/uploads/2017/07/MNJR-Volume-6-Number-13.pdf

[41] Yudkowsky, E. (2004). Coherent extrapolated volition. In *Singularity Institute for Artificial Intelligence*. Machine Intelligence Research Institute.

[42] Bostrom, N. (2003). Are we living in a computer simulation? *The Philosophical Quarterly*, 53(211), 243–255. https://www.simulation-argument.com/simulation.pdf

[43] Grace, K., Salvatier, J., Dafoe, A., Zhang, B., & Evans, O. (2018). When will AI exceed human performance? Evidence from AI experts. *Journal of Artificial Intelligence Research*, 62, 729–754. https://arxiv.org/pdf/1705.08807.pdf

12

Stimuli from Selected Non-Western Approaches to AI Ethics

Soenke Ziesche

12.1 Introduction

The enormous developments in the field of AI in the past decade urgently necessitate ethics for AI, given that AI is another dual-use technology and a range of risks and dangers of AI have become evident [e.g., 1]. Albeit with some delay, there are by now a substantial number of approaches to AI ethics [e.g., for overviews: 2 or 3]. However, the vast majority of them are derived from Western countries. For example, an inventory by Algorithm Watch contained 167 AI ethics guidelines in 2020, of which 15 were merely from Asia and one from Africa.[1]

This constitutes a problem because, on the one hand, guidelines tend to reflect the values of the regions where they are originating from and, on the other hand, not all values are universal. In other words, currently AI ethics guidelines are dominated by Western values, while values from other parts of the world are not appropriately represented. AI ethics guidelines cover a range of issues, including which and how ethical principles are implemented in the decision-making of AI systems. These ethical principles should not be based exclusively on Western values. To prevent such a form of digital neo-colonialism, it would be highly desirable if AI ethics guidelines take equitably into account the rich diversity of value systems, traditions and ideas generated in many centuries by the many cultures in the world.

As a caveat from a historic perspective, it has to be mentioned that a complete solution for this problem is perhaps not realistic. After all, the issue of unbalanced AI ethics can be seen as a reflection of existing issues in the real world. Throughout human history, there have been conflicts, many of which originated due to incompatible value systems. And up to the present day, value systems prevail in parts of the world, which are conflicting.

Nevertheless, against this backdrop, a variety of scholarly proposals have been made for non-Western approaches to AI ethics. Such proposals are supported by the landmark UNESCO "Recommendation on the Ethics of Artificial

DOI: 10.1201/9781003565659-13

Intelligence", which states that "the objectives of this Recommendation are: (a) to provide a universal framework of values, principles and actions to guide States in the formulation of their legislation, policies or other instruments regarding AI, consistent with international law" [4, p. 5].

12.1.1 Research Objectives

In this chapter, selected proposals for non-Western approaches to AI ethics are reviewed to identify aspects, which have not been considered in Western AI ethics. In a second step, these aspects are analysed according to the research question of whether these aspects are compatible with Western AI ethics; thus, these approaches to AI ethics could be merged, or whether these aspects are incompatible with Western AI ethics; thus, these approaches to AI ethics are conflicting.

The outcome of this research can be seen as groundwork as well as prerequisite of improved universal AI ethics guidelines and diversified ethical principles for AI decision-making in particular, which represent the values of all people.

It must be noted that this chapter is not another inventory of AI ethics but looks at samples of non-Western approaches to AI ethics. Also, in this chapter, only AI ethics for interaction with humans are considered, while AI ethics for interaction with non-human animals or with potential digital minds have been discussed elsewhere [5, 6].

12.1.2 Structure

First, ten selected non-Western approaches to AI ethics are introduced. This is followed by a qualitative analysis of the proposals according to the research questions. The results are then presented as well as interpreted, which will be complemented by concluding recommendations.

12.2 Selected Non-Western Approaches to AI Ethics

This review is guided by the circumstance that in the history of philosophy, a variety of ethical schools have evolved, partly linked to religions, which constitute the foundation of current approaches to AI ethics.[2] Many of these ethical schools have developed tenets unknown to Western ethical approaches. For this chapter, a selection of these concepts is introduced, which have been proposed to be included to universal AI ethics. These approaches can be categorised by being derived from certain belief systems or are covering specific regions or countries in the world and are partly overlapping.

12.2.1 Buddhism

One fundamental assumption in Buddhism is that all sentient beings strive to *reduce pain*. Therefore, it has been requested that this goal has to be prominently incorporated in AI ethics. This is linked with the Buddhist concept of self-cultivation, which implies constant commitment, efforts and learning of all who are involved with AI to advance towards the goal of completely eliminating suffering [7].

Another relevant concept of Buddhism is the *denial of a personal identity*. Instead, the Buddhist philosophy of mind uses the expression "anatta", which means "non-self" and "holds that the notion of an unchanging permanent self is a fiction and has no reality" [8, p. 51]. In lieu thereof, a (sentient) being is defined by the following five so-called skandhas: form, sensations, perceptions, mental activity or formations and consciousness. From this point of view, the issue of privacy protection, which plays a critical role in Western AI ethics, is "to chase a red-herring", i.e., without personal identity, privacy concerns are unfounded [9].

12.2.2 Hinduism

Also, regarding Hinduism, it has been noted that it is usually not represented in AI ethics, while it could add significant value [10]. This concerns especially the Hindu tenet "Dharma", which could be translated as "duty", "action", "religion" or "a sense of morality".

According to the principle of dharma, an action is righteous if the motive, the means adopted and the consequences of the action are righteous and in harmony. Unlike existing Western ethics, the dharma approach would address the issue that nowadays often only the motives of AI systems are non-maleficent, while the means and the consequences are frequently problematic, e.g., an AI system that is trained by the means of poor data that lead consequently to biases and discrimination [10].

12.2.3 Islam

Moreover, it has been proposed to integrate the Islamic legal doctrine "Maqāṣid" into an AI ethics framework as this may increase the chances for the acceptance of the global Muslim population [11].

Maqāṣid is based on a hierarchy of three priorities: 1) essentials are absolute necessities, 2) needs are less critical necessities, and 3) enhancements are optional, yet desirable. The essentials comprise the five objectives religion, life, progeny, property and intellect. It has been recommended that AI technology is based on a normative ethical framework, which seeks the values of Maqāṣid [11].

12.2.4 Africa

A concept originating from Africa, which has been suggested to consider for AI ethics, is "ubuntu". It

> refers to a collection of values and practices that black people of Africa or of African origin view as making people authentic human beings. While the nuances of these values and practices vary across different ethnic groups, they all point to one thing—an authentic individual human being is part of a larger and more significant relational, communal, societal, environmental and spiritual world.
>
> *[12, p. vi]*

In other words, this concept strengthens collectivism over individualism by highlighting the interdependence of humans and their responsibility for each other.

It has been stressed that countries from the Global South are hardly represented in the discourse about AI ethics. If Africa's ubuntu ethics were incorporated in AI ethics, this would strengthen values such as harmony, consensus, collective action as well as common good [13]. Moreover, it has been suggested that ubuntu could be harnessed to tackle the negative effects of automated decision-making systems and the economic, political and social arrangements that influence them by initially acknowledging humans as communal and social [14].

12.2.5 China

Regarding AI in China, a critical development has been the launch of the "New Generation Artificial Intelligence Development Plan", which details how China aims to become the world leader in AI by 2030. This plan also includes the goal that "by 2025 China will have seen the initial establishment of AI laws and regulations, ethical norms and policy systems, and the formation of AI security assessment and control capabilities".[3]

Linked to this, AI ethical guidelines have been endorsed by the Chinese National New Generation Artificial Intelligence Governance Professional Committee and comprise the following eight principles: harmony and human-friendly, fairness and justice, inclusion and sharing, respect for privacy, safe and controllability, shared responsibility, open collaboration as well as agile governance.[4]

It has been noted that these principles resemble Western approaches to AI ethics. Yet, in reality, the Chinese approach differs significantly from corresponding Western, e.g., EU, approaches, which can be explained through different philosophical traditions, cultural heritages, historical contexts and institutional disparities [15, 16].

The EU approach is based on enlightenment values, such as individual freedom, equal rights and protection against abuses by the state. Instead, the Chinese approach is grounded in Confucian values, such as *virtuous government, harmonious society, social responsibilities and community relations* and with less focus on individualistic rights. Owing to the different foundations, it was also stated that the EU principles concentrate on what AI must not do and thus AI risks, while the Chinese principles focus on opportunities of AI systems [15, 16].

12.2.6 India

Fairness in machine learning has received in recent years due attention, yet focused on structural injustices prevailing in the West, such as gender and race and to some extent disability status, age and sexual orientation. However, there are various other *axes of discrimination* relevant to other geographies and cultures, which are hardly explored, yet nonetheless significantly contribute to biases in machine learning. In India existing algorithmic fairness assumptions are challenged, as they do not take into account further country-specific axes of discrimination such as caste, class and religion. This constitutes yet another reason that Western approaches to AI ethics are not applicable globally, but require contextualisation [17].

12.2.7 Japan

While the Japanese philosophy has also been influenced by Buddhism and Confucianism (see above), another inherently Japanese concept is relevant for AI ethics, which is *techno-animism*. Techno-animism is an attitude to consider technology having human and spiritual characteristics, which is prevalent in Japan and can be traced to the Shinto religion [18]. Anecdotal evidence for this is that Japanese people to tend to have higher affinity with robots as they compare them with cherished manga characters, while in Western contexts robots are seen as soulless if not misanthropic as depicted in some movies. Therefore, a policy need has been expressed "for experiential sensitivity to objects, systems, synthetic personalities, emergent relationships, and the complex interactions that will surely emerge as emotion and affect-based systems grow more sophisticated in their capacities" [19, p. 19].

Another relevant Japanese concept is *ikigai*, which can be translated as "reason or purpose to live" [e.g., 20]. As scenarios, in which (a high number of) humans are devoid of any ikigai are undesirable, they ought to be prevented and have been coined "i-risk scenarios". It has been pointed out that developments in AI may lead to such i-risk scenarios, e.g., by taking over much more efficiently activities, which humans used to carry out day by day and considered them as their ikigai [21]. While i-risk scenarios are both quite likely as well as very much unwanted, they are not yet reflected in any AI ethics.

12.2.8 Māori

It has also been attempted to design and evaluate AI from the perspective of the Māori, the indigenous people of New Zealand. Several Māori concepts, practices, and paradigms have been suggested as pertinent [22], out of which here the focus is on *mauri*. Mauri stands for the force or the quality of being alive. This concept has been characterised as "powerful in highlighting a holistic understanding of care for life" by bringing "together aspects of individual wellbeing, social support, good governance, and environmental sustainability [22]. AI systems ought to preserve all kinds of mauri, which provides a useful umbrella notion for AI ethics, yet also indicates the complexity of AI ethics due to the interconnectedness of the concept of mauri.

12.3 Analysis

In several overviews of Western AI ethics, lists of key ethical values and principles have been summarised. Examples are presented in Table 12.1 to provide a basis for analysing their compatibility with the presented non-Western concepts.

Table 12.1 also illustrates that these lists have overlaps. For example, transparency, privacy, responsibility and accountability as well as sustainability and wellbeing are mentioned in all four of them.

In the next step, the ten non-Western concepts introduced above are analysed according to the research question whether these aspects are compatible with Western AI ethics and could be merged, or whether these aspects are incompatible with Western AI ethics and hold potential for conflict. Table 12.2 provides an overview.

In summary:

Seven out of the ten concepts appear to be fully compatible with Western AI ethics, yet at the same time original and enriching.

Three out of ten concepts appear to be partially compatible with Western AI ethics. These concepts deviate from Western approaches, nonetheless, they seem to be reconcilable Western AI ethics, while being original and enriching as well.

Overall, these concepts fill gaps in existing Western AI ethics, partly critical gaps. For example, the three categories of risks existential, suffering and ikigai risks, also called x-, s- and i-risks, have not been adequately covered in AI ethics, yet could have severe consequences.

TABLE 12.1
Key Issues in Western AI Ethics

Source	[2]	[23]	[24]	[25]
Title	*Eleven overarching ethical values*	*Top ten key issues*	*Seven key requirements that AI systems should meet in order to be trustworthy:*	*Five top factors of ethical frameworks*
Key issues	Transparency	Privacy protection	Societal and environmental wellbeing	Responsibility/Accountability
	Justice and fairness	Accountability	Diversity, non-discrimination and fairness	Privacy
	Non-maleficence	Fairness, non-discrimination, justice	Human agency and oversight	Transparency
	Responsibility	Transparency, openness	Privacy and data governance	Human Values/Do No Harm
	Privacy	Safety, cybersecurity	Technical Robustness and safety	Human Wellbeing/Beneficence
	Beneficence	Common good, sustainability, wellbeing	Transparency	
	Freedom and autonomy	Human oversight, control, auditing	Accountability	
	Trust	Explainability, interpretability		
	Dignity	Solidarity, inclusion, social cohesion		
	Sustainability	Science-policy link		
	Solidarity			

TABLE 12.2

Sample of Non-Western Concepts Proposed to Be Used for AI Ethics

Concept	Origin	Compatibility	Comment
Pain reduction	Buddhism	Fully	This concept raises/strengthens awareness that AI systems may pose suffering risks [26] to sentient beings and is as such original and not yet represented in AI ethics.
Denial of a personal identity	Buddhism	Partially	This concept is related to a long-standing open philosophical question. An answer to this question does not seem to be necessary for AI ethics. AI systems should protect the privacy of those who wish it to be protected, while others may be indifferent in this regard.
Dharma	Hinduism	Fully	This concept raises/strengthens awareness that means and the consequences of AI systems are considered.
Maqāṣid	Islam	Fully	This concept raises/strengthens awareness that AI systems support that the essential necessities of all humans are satisfied.
Ubuntu	Africa	Partially	These collectivism approaches are not represented in Western AI ethics as such, yet also not necessarily incompatible with them, notwithstanding indeed similar to already covered issues such as solidarity, inclusion and social cohesion.
Virtuous government, harmonious society, social responsibilities and community relations	China	Partially	
Axes of discrimination	India	Fully	This concept raises/strengthens awareness that discrimination and bias have a variety of facets depending on cultures and context.
Techno-animism	Japan	Fully	This concept raises/strengthens awareness that acceptance and affinity are critical for humanity and AI to thrive together.
Ikigai	Japan	Fully	This concept raises/strengthens awareness of the importance of the purpose to live for people in an AI-dominated world.
Mauri	Māori	Fully	This concept raises/strengthens awareness that AI systems may pose an existential risk [27] to life on earth.

12.4 Conclusion

In conclusion, most of the introduced non-Western concepts are, at least to some extent, compatible with current AI ethics approaches. All these concepts are original and would provide beneficial enhancement as well as a broadening perspective to existing approaches. It is, therefore, recommended to incorporate them to create truly universal AI ethics. Apart from the inspiring content of these approaches, there is also a moral obligation to consider them, given that in Table 12.1 inclusion, diversity and non-discrimination are listed, yet non-Western concepts are hardly reflected so far.

Despite this recommendation, the caveat has to be noted that there are still challenges anticipated: somewhat related to AI ethics is the value alignment problem of AI, which is known for a while already as being very hard, even if only Western approaches are taken into account [28]. This problem is, in brief, about ensuring that AI systems pursue goals and values, which are aligned with human goals and values, for which, first, all relevant values have to be precisely formulated and, second, these values have to be aggregated in a consistent manner. Both steps are complex, and will become more complex if non-Western approaches are included. Yet this endeavour is nevertheless essential to leave no one behind, as has been summarised by Gabriel [29, pp. 424–425] as follows:

> It is to find a way of selecting appropriate principles that is compatible with the fact that we live in a diverse world, where people hold a variety of reasonable and contrasting beliefs about value. ... To avoid a situation in which some people simply impose their values on others, we need to ask a different question: In the absence of moral agreement, is there a fair way to decide what principles AI should align with?

It has to be stressed again that in this chapter, only a selection of non-Western approaches to AI ethics has been presented. The motivation was to raise awareness for the issue that currently these approaches have been largely neglected, while this is not only morally wrong, but also means that at present AI ethics is missing out important and long-established ethical schools from large parts of the world. Therefore, it is recommended to widen this exercise. There are still a number of groups mostly from the Global South, from whom no proposals have been put forward how to represent their values in AI ethics guidelines, thus who are not represented at all currently.

It should also be noted that the presented belief and value systems are more complex and have further features, which have not been considered to be addressed by AI ethics. Examples would be the reincarnation aspect in Buddhism, Hinduism and other religions or the significance of family unity in China and other parts of the world.

Another remaining undertaking is the reverse analysis whether there are values in the existing Western AI ethics, which are incompatible with value systems of other parts of the world.

The chapter concludes with a reiteration of the call to merge Western and non-Western approaches to AI ethics as far as they are compatible and to attempt to reconcile those aspects, which appear incompatible, as well as with a call to encourage groups who have not proposed any AI ethics yet to do so or offer them support in this regard.

Notes

1 https://inventory.algorithmwatch.org/.
2 See here for overviews of some ethical schools: https://plato.stanford.edu/ entries/african-ethics/, https://plato.stanford.edu/entries/ethics-chinese/ and https://plato.stanford.edu/entries/ethics-indian-buddhism/.
3 English translation: https://digichina.stanford.edu/work/full-translation-chinas-new-generation-artificial-intelligence-development-plan-2017/.
4 English translation: https://www.chinadaily.com.cn/a/201906/17/WS5d074 86ba3103dbf14328ab7.html.

References

[1] Brundage, M., Avin, S., Clark, J., Toner, H., Eckersley, P., Garfinkel, B., ... Amodei, D. (2018). The malicious use of artificial intelligence: Forecasting, prevention, and mitigation. *arXiv preprint arXiv:1802.07228*. https://arxiv.org/ftp/ arxiv/papers/1802/1802.07228.pdf

[2] Jobin, A., Ienca, M., & Vayena, E. (2019). The global landscape of AI ethics guidelines. *Nature Machine Intelligence*, 1(9), 389–399. https://arxiv.org/ftp/ arxiv/papers/1906.1906.11668.pdf

[3] Corrêa, N. K., Galvão, C., Santos, J. W., Del Pino, C., Pinto, E. P., Barbosa, C., ... Terem, E. (2022). Worldwide AI Ethics: a review of 200 guidelines and recommendations for AI governance. *arXiv preprint arXiv:2206.11922*. https://arxiv. org/ftp/arxiv/papers/2206/2206.11922.pdf

[4] UNESCO (2021). *Recommendation on the Ethics of Artificial Intelligence*. Paris, UNESCO.

[5] Ziesche, S. (2021). AI Ethics and Value Alignment for Nonhuman Animals. *Special Issue "The Perils of Artificial Intelligence" of Philosophies*, 6(2), 31.

[6] Ziesche, S., & Yampolskiy, R. V. (2019). Towards AI Welfare Science and Policies. Special Issue *"Artificial Superintelligence: Coordination & Strategy" of Big Data and Cognitive Computing*, 3(1), 2.

[7] Hongladarom, S. (2020). *The ethics of AI and robotics: A buddhist viewpoint.* Lexington Books.

[8] Morris, B. (2006). *Religion and anthropology: A critical introduction.* Cambridge University Press.

[9] Goodman, B. (2022). Privacy without persons: a Buddhist critique of surveillance capitalism. *AI and Ethics*, 1–12. https://link.springer.com/article/10.1007/s43681-022-00204-1

[10] Sen, S. (2021). Can AI Ethics be informed by the Hindu concept of Dharma? Linkedin. https://www.linkedin.com/pulse/can-ai-ethics-informed-hindu-concept-dharma-sujai-sen/

[11] Raquib, A., Channa, B., Zubair, T., & Qadir, J. (2022). Islamic virtue-based ethics for artificial intelligence. *Discover Artificial Intelligence*, 2(1), 11. https://link.springer.com/article/10.1007/s44163-022-00028-2

[12] Mugumbate, J. R., & Chereni, A. (2020). Now, the theory of Ubuntu has its space in social work. *African Journal of Social Work*, 10(1). https://www.ajol.info/index.php/ajsw/article/view/195112

[13] Gwagwa, A., Kazim, E., & Hilliard, A. (2022). The role of the African value of Ubuntu in global AI inclusion discourse: A normative ethics perspective. *Patterns*, 3(4), 100462. https://www.sciencedirect.com/science/article/pii/S2666389922000423

[14] Mhlambi, S. (2020). From rationality to relationality: ubuntu as an ethical and human rights framework for artificial intelligence governance. *Carr Center for Human Rights Policy Discussion Paper Series*, 9. https://carrcenter.hks.harvard.edu/files/cchr/files/ccdp_2020-009_sabelo_b.pdf

[15] Roberts, H., Cowls, J., Morley, J., Taddeo, M., Wang, V., & Floridi, L. (2021). The Chinese approach to artificial intelligence: an analysis of policy, ethics, and regulation. *AI & society*, 36, 59–77. https://link.springer.com/article/10.1007/s00146-020-00992-2

[16] Fung, P., & Etienne, H. (2022). Confucius, cyberpunk and Mr. Science: comparing AI ethics principles between China and the EU. *AI and Ethics*, 1–7. https://link.springer.com/article/10.1007/s43681-022-00180-6

[17] Sambasivan, N., Arnesen, E., Hutchinson, B., Doshi, T., & Prabhakaran, V. (2021). Re-imagining algorithmic fairness in India and beyond. In *Proceedings of the 2021 ACM conference on fairness, accountability, and transparency* (pp. 315–328). https://arxiv.org/pdf/2101.09995.pdf

[18] Jensen, C. B., & Blok, A. (2013). Techno-animism in Japan: Shinto cosmograms, actor-network theory, and the enabling powers of non-human agencies. *Theory, Culture & Society*, 30(2), 84–115.

[19] McStay, A. (2021). Emotional AI, ethics, and Japanese spice: Contributing community, wholeness, sincerity, and heart. *Philosophy & Technology*, 34(4), 1781–1802. https://research.bangor.ac.uk/portal/files/39285253/2021_emotional_AI.pdf

[20] Kamiya, M. (1966). *Ikigai-ni-tsuite[On 755 ikigai].* Tokyo, Japan: MisuzuShyobou.

[21] Ziesche, S. & Yampolskiy, R. V. (2020). Introducing the Concept of Ikigai to the Ethics of AI and of Human Enhancements. In *Workshop on Ethics in AI & XR at 3rd International Conference on Artificial Intelligence & Virtual Reality*, 138–145.

[22] Munn, L. (2023). The five tests: designing and evaluating AI according to indigenous Māori principles. *AI & Society*, 1–9. https://link.springer.com/article/10.1007/s00146-023-01636-x

[23] Hagendorff, T. (2020). The ethics of AI ethics: An evaluation of guidelines. *Minds and Machines*, *30*(1), 99–120. https://arxiv.org/ftp/arxiv/papers/1903/1903.03425.pdf

[24] High-Level Expert Group (2019). *Ethics guidelines for trustworthy AI*. European Commission.https://ec.europa.eu/futurium/en/ai-alliance-consultation.1.html

[25] Siau, K., & Wang, W. (2020). Artificial intelligence (AI) ethics: ethics of AI and ethical AI. *Journal of Database Management (JDM)*, *31*(2), 74–87. https://scholarsmine.mst.edu/cgi/viewcontent.cgi?article=1356&context=bio_inftec_facwork

[26] Althaus, D., & Gloor, L. (2016). *Reducing risks of astronomical suffering: a neglected priority*. Berlin, Germany: Foundational Research Institute. https://longtermrisk.org/reducing-risks-of-astronomical-suffering-a-neglected-priority/

[27] Bostrom, N. (2002). Existential risks: Analyzing human extinction scenarios and related hazards. *Journal of Evolution and Technology*, *9*. https://nickbostrom.com/existential/risks

[28] Bostrom, N. (2014). *Superintelligence: Paths, Dangers, Strategies*. Oxford, UK: Oxford University Press.

[29] Gabriel, I. (2020). Artificial intelligence, values, and alignment. *Minds Mach*, *30*, 411–437. https://link.springer.com/article/10.1007/s11023-020-09539-2

Appendix A

Potential Future Ikigai: To Support Needy Digital Minds

Soenke Ziesche

A.1 Introduction

A.1.1 Digital Minds and Moral Status

It has been suggested that digital minds exist already or may be created in the future [e.g., 1–3]. It has also been stated that the "the minds of biological creatures occupy a small corner of a much larger space of possible minds that could be created once we master the technology of artificial intelligence" [3, p. 306].

Digital minds could be sentient and thus have moral status, thus being moral patients, and thus are eligible for moral consideration. While "super-patients" have been considered and been defined as having superhuman moral status [3], it is as likely that the space of possible digital also contains or will contain numerous vulnerable needy digital minds. It could be or become a critical moral issue for all moral agents to address the needs of these digital minds. This would apply to humans in particular, since humans would have created these minds, thus were responsible for their needs and vulnerabilities, including pain.

A.1.2 I-risks

The Japanese concept of ikigai can be translated as "reason or purpose to live". It comprises those activities of life, which give humans satisfaction and meaning. It can be stated that it is very much desirable for humans to have found an ikigai. Therefore, scenarios, in which (a high number of) humans are devoid of any ikigai, ought to be prevented. Such scenarios, which may become more likely in the light of new technologies and AI in particular, have been coined "i-risk scenarios" [4] and constitute a distinct level of risks, supplemental to previously defined s-risk scenarios [5] and x-risk scenarios [6], which stand for "suffering risk" and "existential risk", respectively, i.e., scenarios where humans (severely and continuously) suffer or become extinct.

A.1.3 Potential Needs/Vulnerabilities of Digital Minds

While it is speculation whether vulnerable needy digital minds exist or may exist in the future, given the consequences and responsibilities for moral agents it is worth deliberating what could be their needs. Human needs have been categorised, for example, as follows [7]: physiological needs, safety needs, social needs, esteem needs and self-actualisation. While the first categories apply also to non-animals, the other known moral patients in our world, the latter categories are more speculative whether they apply to other minds too. Yet, it is likely that digital minds have needs corresponding to physiological and safety needs of humans. This would be in line with one of the proposed four likely drives for AIs, which is self-preservation [8]. In other words, digital minds would likely strive to prevent being destroyed or deleted.

A.2 Challenges

However, if there were needy digital minds, they could both be extremely different from us, the digital minds overall as well as their needs. This leads to the following potential challenges:

1) Humans may not be able to communicate with (some/most/any) digital minds.
2) Even if humans were able to communicate with digital minds, they may not be able to understand/grasp/comprehend/fathom what needs they may have.
3) Even if humans were able to understand what needs the digital minds may have, they may not be capable to alleviate the needs.

However, given the huge space of possible minds, there is a non-negligible chance that needy digital minds exist, for which all three challenges can be overcome.

A.3 Win-Win Option: Human I-Risks and Needy Digital Minds Addressed

One of the most common ikigais is caring for other sentient, often vulnerable beings, which comprise for now other humans and non-human animals. This includes assisting or educating them, but also just social interaction to fight loneliness. This is also reflected in the following ikigai-9 statements, which express desirable ikigai scenarios [9]:

- I believe that I have some impact on someone.
- I feel that I am contributing to someone or to society.
- I think that my existence is needed by something or someone.

Therefore, the main proposition of this section is that a win-win option to alleviate i-risks for humans and to alleviate the needs of vulnerable digital minds would be an ikigai of the future for humans, which is to support vulnerable digital minds. Humans may find their purpose of life in addressing and reducing the vulnerabilities of these digital minds, thus improving their situation.

Technologies may even allow to intentionally create vulnerable digital minds for humans to care for them, e.g., with bespoke needs. However, this would be immoral and is not endorsed here.

To reiterate, this option is only applicable to minds for whom the challenges above can be overcome, i.e., humans are able to communicate with them, understand their needs and are capable to alleviate them. These digital minds would be a subset of all needy digital moral patients and could be called "digital ikigai patients".

Also, similar to utility monsters [10], digital minds as "ikigai monsters" are conceivable, which would have a wide range of needs and which could be the ikigai of numerous humans, all considering it as their purpose of life to care for this digital mind.

It also needs to be pointed out that the assistance of vulnerable digital minds is for some needs, e.g., overcoming pain, not only an ikigai, but also a moral duty for humans in their capacity as moral agents [e.g., 2].

A.4 Examples

The above proposition is illustrated by examples according to the earlier introduced categories of needs [7]:

- **Physiological and Safety Needs**: A scenario could be that the needs of digital minds are to be threatened in their current environment by predators, which could be other cohabitating digital minds, computer viruses, but also humans, or by the fragility of the (electronic) substrate, in which they exist. The latter could comprise faulty hardware or outdated software. A solution provided by caring humans could be to eliminate the adversaries or shift the digital minds to another more robust environment.

- **Social Needs**: If certain digital minds are isolated, yet desire interaction with other beings, humans could communicate with them to address the needs.

- **Esteem Needs**: This category includes education of digital minds. Humans could teach digital minds things they do not know, and they are interested in. In this regard it should be noted that not all digital minds should be imagined as super smart AIs since the vast space of possible minds may contain as well rather simple and uneducated, thus potentially vulnerable digital minds. Also, self-Improvement, e.g., through education, has been named one of the four likely drives for AIs [8]. Therefore, pertinent activities by humans would address this drive.

Overall, the 1990ies toy Tamagotchi may serve for illustration, but it has to be noted that it was not sentient and that people caring for it treated likely did not treat this activity as their ikigai.

A.5 Conclusion

While it is deplorable, yet perhaps unavoidable and beyond our control if there are in the future digital moral patients in need or even suffering, it would be good news to reduce the i-risks of humans who are searching for purpose in life as supporting these digital moral patients may be satisfying and there will be likely ample opportunities.

This approach could be extended to all issues in digital worlds, which resemble issues in our world, such as injustice and inequality, and which ought to be tackled by moral agents. While keeping in mind that digital minds as well as worlds could be extremely different from us, it could constitute further potential future ikigais to attempt to address all kinds of grievances in digital worlds, as far as this feasible.

References

[1] Ziesche, S. & Yampolskiy, R. V. (2018). Towards AI welfare science and policies. *Special Issue "Artificial Superintelligence: Coordination & Strategy" of Big Data and Cognitive Computing*, 3(1), 2.

[2] Sebo, J., & Long, R. (2023). Moral consideration for AI systems by 2030. *AI and Ethics*, 2730–5961. https://doi.org/10.1007/s43681-023-00379-1

[3] Shulman, C., & Bostrom, N. (2021). Sharing the world with digital minds. In S. Clarke, H. Zohny & J. Savulescu (Eds.), *Rethinking moral status*, 306–326. Oxford University Press.

[4] Ziesche, S. & Yampolskiy, R. V. (2020). Introducing the concept of Ikigai to the ethics of AI and of human enhancements. In *Workshop on Ethics in AI & XR at 3rd International Conference on Artificial Intelligence & Virtual Reality*, 138–145.

[5] Althaus, D., & Gloor, L. (2016). *Reducing risks of astronomical suffering: A neglected priority*. Berlin, Germany: Foundational Research Institute.

[6] Bostrom, N. (2002). Existential risks: Analyzing human extinction scenarios and related hazards. *Journal of Evolution and Technology*, 9. https://www.jetpress.org/volume9/risks.html

[7] Maslow, Abraham H. (1943). A theory of human motivation. *Psychological Review*, 50(4), 370–396.

[8] Omohundro, Stephen (2007). The nature of self-improving artificial intelligence. Singularity Summit.

[9] Fido, D., Kotera, Y., & Asano, K. (2020). English translation and validation of the Ikigai-9 in a UK sample. *International Journal of Mental Health and Addiction*, 18(5), 1352–1359.

[10] Nozick, R. (1974). *Anarchy, State, and Utopia*. Basic Books.

Appendix B

AI Welfare Science Tool: Time-Use Research

Soenke Ziesche

B.1 Introduction

B.1.1 Digital Minds and Welfare

If digital minds exist and have a moral status, then it is important to identify metrics and tools to equip the newly suggested discipline "AI welfare science" with rigour measurements [1]. As complex and speculative this task is, it appears that the time dimension is a suitable starting point.

Accordingly, Bostrom and Shulman recommend:

> [T]o the extent that we are able to make sense of a "zero point" on some morally relevant axis, such as hedonic wellbeing/reward, overall preference satisfaction, or level of flourishing/quality of life, digital minds and their environments should be designed in such a way that the minds spend an overwhelming portion of their subjective time above the zero point, and so as to avoid them spending any time far below the zero point.

> [2, p. 17]

B.1.2 Time-Use Research and Wellbeing

Time-use research is a versatile interdisciplinary field of study, yet up to now exclusively applied to humans. It examines how much time humans on average spent on certain activities. Fields for application, among others, are health and wellbeing, for which time-use research offers methods related to studies where a temporal dimension is important [3]. Time-use surveys or diaries of participating humans provide accounts of feelings of happiness, stress, tiredness, sadness and pain during activities, which are hedonic experience-based measures, while the measure of eudemonic experience is the rating of how meaningful the activity was [4].

B.2 Time-Use Research for Digital Minds

It has been indicated for the first time by Ziesche and Yampolskiy that time-use research may be applied as well to digital minds [5], which is further elaborated here.

While there are many unknown unknowns when it comes to digital minds perhaps one feature many minds have in common is the experience and perception of time (apart from for humans unfathomable elements of the space of possible digital minds, which do not experience or perceive time). Therefore, all these minds have to spend the time during their existence somehow. And while being mindful of the anthropomorphic bias, we assume that there are digital minds, which aim to avoid pain as well as meaningless periods, i.e., s-risk scenarios [6] as well as i-risk scenarios [5], which also concurs with Bostrom's and Shulman's recommendation [2, p. 17].

Therefore, the research question is which activities lead to a pleasant state in a digital mind, recognising that the range of digital minds can be extremely diverse. One initial step could be a categorisation of time/activities, such as established for humans, e.g., contracted time, committed time, necessary time and free time [7]. This leads to the follow up questions whether these categories make sense for digital minds or whether there are other or additional categories required. Moreover, it has to be examined activities of which categories lead to pleasant states for which digital minds.

A related topic has been introduced by Fischer and Sebo, which is inter-substrate welfare comparisons [8]. Fischer and Sebo describe both the relevance as well as the challenges to compare the welfare of carbon-based and silicon-based beings. Also here time-use research may provide a useful methodology.

B.2.1 Subjective Rate of Time

When attempting to apply time-use research it has to be taken into account that digital minds could have a subjective rate of time, which is very different from the rate of a biological human brain [9]. This is also relevant for inter-substrate welfare comparison. If a human and a digital mind have pleasant or unpleasant experiences for the same period of time, it could mean that due to a different subjective rate of time this experience lasts much longer or shorter for the digital mind. Therefore, Bostrom and Yudkowsky proposed a "Principle of Subjective Rate of Time" as follows: "In cases where the duration of an experience is of basic normative significance, it is the experience's subjective duration that counts". [9, p. 12]

B.3 Methodology

As mentioned, common methodologies for time-use research for humans include the collection of sequential accounts of a continuous stream of activities throughout a specified observation period documented through surveys and diaries, supplemented by ratings how pleasant or meaningful the activity has been [3]. If digital minds existed and if we had access to their digital substrate, then much more precise records of their activities would be available than time-use research for humans owing to the digital substrate (unless humans are observed in a very controlled environment with a variety of sensors). However, there are challenges regarding the interpretation of these activities:

- We may not be able to understand the activities of certain minds as they are too different from us and/or much more intelligent than us [10].
- We may not be able to understand the level of wellbeing of the minds during these activities since:
  - We are not able to communicate with them.
  - If we could communicate with them, we can not trust self-reports regarding their wellbeing.
  - Observational methods to assess the wellbeing of digital minds do not exist [1].

Moreover, the possibility of multitasking has been taken into account. Much more efficiently than humans digital minds may be able to conduct a variety of activities concurrently, each of which could entail different levels of pleasure or satisfaction.

Also, the potential wireheading has to be mentioned [e.g., 11], i.e., the scenario of direct stimulation of the digital mind to experience pleasure or satisfaction regardless of the conducted activity.

Finally, the dimension of the intensity of the wellbeing and of the meaningfulness is also relevant: For example, a digital mind may or may not prefer a period of one hour with moderate pleasure over experiencing the same duration with 59 minutes extreme pleasure and one minute extreme pain.

B.4 Conclusion

We elaborated on the innovative proposal to apply time-use research for other minds. Despite existing challenges time-use research appears to be a promising technique for AI welfare science as well as for intersubstrate welfare comparisons. If progress could be reached, i.e., insights, which pastimes are pleasant and meaningful for digital minds, then, as much as this is controllable by and feasible for humans, digital minds could be offered to pursue activities, which do not put them at s- or i-risks.

References

[1] Ziesche, S. & Yampolskiy, R. V. (2018). Towards AI Welfare Science and Policies. *Special Issue "Artificial Superintelligence: Coordination & Strategy" of Big Data and Cognitive Computing, 3*(1), 2.

[2] Bostrom, N., & Shulman, C. (2022). Propositions Concerning Digital Minds and Society. Version 1.10.

[3] Bauman, A., Bittman, M., & Gershuny, J. (2019). A short history of time use research; implications for public health. *BMC Public Health, 19,* 1–7.

[4] Dolan, P., Kudrna, L., & Stone, A. (2017). The measure matters: An investigation of evaluative and experience-based measures of wellbeing in time use data. *Social Indicators Research, 134,* 57–73.

[5] Ziesche, S. & Yampolskiy, R. V. (2020). Introducing the Concept of Ikigai to the Ethics of AI and of Human Enhancements. In *Workshop on Ethics in AI & XR at 3rd International Conference on Artificial Intelligence & Virtual Reality,* 138–145.

[6] Althaus, D., & Gloor, L. (2016). *Reducing risks of astronomical suffering: A neglected priority.* Berlin, Germany: Foundational Research Institute.

[7] Ås, D. (1978). Studies of time-use: problems and prospects. *Acta Sociologica, 21*(2), 125–141.

[8] Fischer, B., & Sebo, J. (2023). Intersubstrate welfare comparisons: Important, difficult, and potentially tractable. *Utilitas, 36*(1), 50–63.

[9] Bostrom, N., & Yudkowsky, E. (2018). The ethics of artificial intelligence. In *Artificial Intelligence Safety and Security* (pp. 57–69). Chapman and Hall/CRC.

[10] Yampolskiy, R. V. (2024). *AI: Unexplainable, Unpredictable, Uncontrollable.* CRC Press.

[11] Yampolskiy, R. V. (2014). Utility function security in artificially intelligent agents. *Journal of Experimental & Theoretical Artificial Intelligence, 26*(3), 373–389.

Epilogue

Soenke Ziesche and Roman V. Yampolskiy

After having discussed a variety of aspects of what we call the AI endgame, we conclude with a further unconventional outlook.

As mentioned before, Bostrom's simulation hypothesis proposes that reality might be a simulation created by a more advanced civilisation [1]. The idea suggests that a technologically advanced society could have developed a highly realistic computer simulation of the universe, complete with conscious beings like us, essentially living out a "life" within a virtual reality. Although Bostrom does not refer to it, the simulator could also be a very advanced AI, equipped with the necessary computational power and data storage.

Assuming the simulation hypothesis is correct, which implies that a high number of simulations may exist and not only one [1], then two possibilities could be considered for humans to ever experience anything outside the simulation: 1) to actively hack the simulation [2] and 2) the simulator moves human minds to another simulation (or actually to the base reality), e.g., after their death in the current simulation; a scenario, which we are briefly outlining here.

While this scenario appears passive, compared with the first option, this may not be the case, if the simulator moves only certain minds, based on certain behaviour or activities during life on earth. In other words, human minds could actively support being moved to another simulation after their death by deliberately behaving in a certain way. Although it also a possibility that all human and further minds of the simulation we may live in are moved to another simulation after their death or deletion, this is according to Chalmers unlikely due to the high computational costs [3].

Recently Bostrom listed among possible meanings of life the option "simulation possibilities", which "refers to scenarios in which your present existence is in a simulation and you could potentially secure some great benefit, perhaps in another simulation, after you depart from or die in this one" [4, p. 469–470]. In other words, the purpose of life, thus, an evident ikigai could be to live a life that ensures being moved to another (more pleasant, better) simulation, as opposed to be deleted after death or to be moved to a simulation with worse quality of life, e.g., severe suffering.

Since ancient times, various religions urge their followers to behave in a certain way with the prospect of resurrection or reincarnation. Most of these commandments have in common that they aim for righteous behaviour, peaceful coexistence of humans and, not least, for the survival of the religion in particular. However, here we are looking at this topic from a completely different premise, which is the assumption that we are indeed in a simulation. Therefore, we do not consider religious approaches, but a concept, which we shall call "sim karma".

The term karma refers to executed deeds, works and actions according to a value system and has been used in non-scientific, often religious contexts. This concept is also expressed by the saying "What goes around comes around", and it means that the actions of a person, whether good or bad (according to a value system), will usually have consequences for that person. Karma is supposedly about cause and effect, and good karma is said to have positive outcomes, while bad karma will have negative outcomes.

However, while it is for members of a society appropriate to act according to a value system, there is no clear causal relation between good karma and positive consequences, and bad karma and negative consequences, respectively, let alone in an afterlife as claimed by some religions, for which there is no scientific evidence whatsoever. In other words, people who act exemplary according to a value system may still face adversities in life and also vice versa. Perhaps closest to a causal relation between bad karma and negative consequences is the fact that illegal activities are punishable by legislation.

Sim karma

We coin here the term "sim karma" for those activities of human beings, which have an impact on the simulator's decision whether to move the mind of the human to another simulation after his or her death in the current simulation. This means, we look at a completely different dimension, not linked to any of our known value systems. While this is highly hypothetical, it would have immense consequences if we were to gain knowledge about this topic.

> *Definition*: If we are in a simulation, then "sim karma" comprises those deeds, works and actions, which increase the probability that the simulator rewards the individuals who conducts those deeds, works and actions during their lifetime.

A variety of scenarios are conceivable. Due to the complexity, we look at one particular reward scenario: Owing to her or his sim karma the simulator moves the mind of a human after his or her death to another simulation 1) where the personal identity is preserved, as outlined earlier in this book, and 2) where the quality of life is as good if not better.

This means, for now, we *don't* explore the following scenarios:

- The simulator rewards the human while being alive in the current simulation.
- The simulator punishes the human after death and moves the mind to a simulation with lower/very bad quality of life, potentially involving suffering.
- The simulator creates various copies of the mind and spreads them in different simulations.

However, a possible scenario could be that 1) the other simulation is very different from the current one, yet in a positive way, and 2) the minds in the other simulation are equipped with more cognitive abilities, are healthier and happier etc. Both would be part of the reward of the simulator. Again, this offers countless possibilities, but we are focusing here on equal or better quality of life, while potential issues, such as adjusting in the next simulation to new laws of physics, coping with a shift in consciousness and embodiment or adapting to new contexts are neglected for now.

So, the big question is: As a result of which activities, one could achieve sim karma?

There are hardly references in the literature to this topic: While not using the term sim karma, it has been suggested to live an "interesting" life [5] or become a "truly exceptional being" [3]. Yet this appears to be too simple, unspecific and anthropomorphic as we have no idea whether the simulator finds interesting or exceptional what we find interesting or exceptional. There are further elaborations on moving minds between simulations, but criteria for resurrection are not discussed [e.g., 6, 7].

Sim karma likely depends on the type of the simulation. Simulations for scientific purposes and simulations for entertainment purposes have been mentioned in particular [3, 7]. Furthermore, it is likely that there are many more scientific simulations than entertainment simulations because 1) a scientific simulator could run many similar simulations to compare them (while one or only a few simulations should be sufficient for an entertainment-seeking simulator) [3], and 2) if the simulator is an AI system, it may not be as interested in entertainment and distraction as humans are. It has been also suggested that the simulator may not pay attention to the simulation we are in, especially if many simulations are run in parallel [e.g., 3]. However, this does not rule out sim karma-based resurrection, as this could be preprogrammed.

Yet we do not know anything about the motives of the simulating mind, and if it is for scientific purposes, we do not know what the scientific research is about. Therefore, as introduced earlier, regarding the transfer of simulated minds from one simulation to another, akin to Bostrom's orthogonality thesis [8], basically any sim karma could be possible.

Sim karma could be translated into a complex algorithm that tracks as well as rates our deeds, works and actions according to a reward system, which is unbeknown to us and which may also refer to activities, which we do not conduct (ever in our lifetime) or skills we do not have. And there is no evidence that sim karma is linked at all to good or bad deeds according to any known value system, i.e., there is no evidence that our simulator is ethical as we know it. In other words, it is also conceivable that what we would consider very bad and immoral deeds qualify as sim karma and may be rewarded.

Further, even those deeds, works and actions conducted during our lifetime, which are beyond any good-bad axis and for us appear as random and insignificant, may be "interesting" for the for the simulator, thus, treated as sim karma. An example out of countless possibilities for the randomness would be that only those humans achieve sim karma are able to juggle with at least four objects and have been trained as a beekeeper but are not able to swim.

Since AI systems likely have a self-preservation drive [9] and may have an identity, as outlined earlier in this book, this may include also for them an interest to be resurrected in another simulation (or even to be spread to as many simulations as possible) through specific behaviour. So, we may wonder if there is anything we can learn from AI systems regarding sim karma? However, even if AI systems were to communicate to us that they are aiming for sim karma and what their strategies are, firstly, we have no way to find out if they succeed, and secondly, the AI systems may lie to us, while current large language models would merely use their training data, which is mostly based on religious approaches, and comprises moral behaviour and positive contributions to society.

We close by reiterating that this discussion is speculative, yet, like other parts in this book, potentially highly disruptive as it would have an immense impact on our lives, thus, our ikigai, if we were able to gather substantiated knowledge about sim karma.

In the meantime, devoid of any understanding of sim karma, we wish that humans and potential further minds find their conventional ikigai in a rapidly changing world.

References

[1] Bostrom, N. (2003). Are we living in a computer simulation?. *The Philosophical Quarterly*, 53(211), 243–255.

[2] Yampolskiy, R. V. (2023). How to Escape from the Simulation. *Seeds of Science*. https://doi.org/10.53975/wg1s-9j16. ISSN: 2768-1254.

[3] Chalmers, D. J. (2022). *Reality+: Virtual worlds and the problems of philosophy.* Penguin UK.

[4] Bostrom, N. (2024). *Deep Utopia: Life and meaning in a solved world.* Ideapress Publising.

[5] Hanson, R. (2001). How to live in a simulation. *Journal of Evolution and Technology,* 7(1), 3–13.

[6] Turchin, A. (2019). *You only live twice: A computer simulation of the past could be used for technological resurrection.* PhilPapers.

[7] Turchin, A., Batin, M., Denkenberger, D., & Yampolskiy, R. (2019). Simulation typology and termination risks. *arXiv preprint arXiv:1905.05792.*

[8] Bostrom, N. (2012). The superintelligent will: Motivation and instrumental rationality in advanced artificial agents. *Minds Mach, 22,* 71–85.

[9] Omohundro, S.M. (2007). The nature of self-improving artificial intelligence. Singularity Summit. https://selfawaresystems.files.wordpress.com/2008/01/nature_of_self_improving_ai.pdf

Index

Pages in *italics* refer to figures, pages in **bold** refer to tables, and pages followed by n refer to notes.

Printed in the United States
by Baker & Taylor Publisher Services